THE UNITED STATES OF AMERICA

NATIONAL REPORT

FOR THE

CONVENTION ON NUCLEAR SAFETY

SEPTEMBER 2001

U.S. NUCLEAR REGULATORY COMMISSION

WASHINGTON DC 20555-0001

ABSTRACT

The U.S. Nuclear Regulatory Commission has updated the *U.S. National Report for the Convention on Nuclear Safety.* This report will be submitted for peer review at the second review meeting of the Convention on Nuclear Safety at the International Atomic Energy Agency in April 2002. The scope of the report is limited to the safety of land-based civil nuclear power plants. The report demonstrates how the U.S. Government meets the main objective of the Convention -- to achieve and maintain a high level of nuclear safety worldwide. The report also shows that the U.S. Government meets the obligations of each of the articles established by the Convention. Specifically, these articles address the safety of existing nuclear installations, the legislative and regulatory framework, the regulatory body, responsibility of the licensee, priority to safety, financial and human resources, human factors, quality assurance, assessment and verification of safety, radiation protection, emergency preparedness, siting, design, construction, and operation.

TABLE OF CONTENTS

TABLE OF CONTENTS (CONT.)

TABLE OF CONTENTS (CONT.)

EXECUTIVE SUMMARY

The U.S. Nuclear Regulatory Commission (NRC) has updated the *U.S. National Report for the Convention on Nuclear Safety,* which will be submitted for peer review at the second review meeting of the Convention in April 2002. The updated report follows the same format as the original U.S. National Report, which was submitted in 1998, and the scope of the report remains limited to the safety of land-based civil nuclear power plants. The report demonstrates how the U.S. Government meets the main objective of the Convention -- to achieve and maintain a high level of nuclear safety worldwide by enhancing national measures and international cooperation. The report also shows that the U.S. Government meets the obligations of each of the articles established by the Convention. Specifically, these articles address the safety of existing nuclear installations, the legislative and regulatory framework, the regulatory body, responsibility of the licensee, priority to safety, financial and human resources, human factors, quality assurance, assessment and verification of safety, radiation protection, emergency preparedness, siting, design, construction, and operation.

The updated report discusses the national nuclear programs that are used to meet the obligations of the Convention, the main current nuclear safety initiatives, and significant regulatory accomplishments since the original report was written in 1998. In addition, the report highlights the NRC's major objectives – to maintain plant safety, to improve the effectiveness, efficiency, and realism of NRC activities and decisions, to increase public confidence, and to reduce unnecessary regulatory burden – and describes many activities that the NRC has initiated to meet those objectives.

The NRC uses a number of national programs and processes to ensure that plant safety is maintained and to meet the obligations of the Convention on Nuclear Safety. These are a well-established licensing process; the newly revised process for reactor oversight; the Accident Sequence Precursor Program; the Program for Resolving Generic Safety Issues; programs for rulemaking; for decommissioning; for regulatory research; for public participation; and for handling petitions, allegations, and differing professional views and opinions.

The main nuclear safety initiatives that the NRC and its licensees are carrying out in 2001 are the development of risk-informed regulatory programs, implementation of a more objective, transparent, and risk-informed reactor oversight process, assessment of the implications of industry restructuring and consolidation, enhancement of the regulatory infrastructure associated with license renewal and pursuit of extensive research on plant aging mechanisms, and preparation for potential licensing of new reactors.

Benefitting from over 40 years of operational experience and steadily improving performance by licensees, the NRC has changed the way that it regulates, mainly by developing a more risk-informed and performance-based regulatory approach. To encourage a sustained high level of safety performance of U.S. nuclear plants, the NRC has improved its most important oversight processes to incorporate risk insights from quantitative risk analysis. The NRC is also continuing efforts to revise regulations to focus requirements on plant programs and activities that are most risk significant. The move toward risk-informing the current regulations and processes marks perhaps the most significant changes taking place at the NRC.

The NRC recently phased out its previous oversight process for nuclear plants, which was considered by many stakeholders to be subjective, unpredictable, and not well understood. The new process, known as the Revised Reactor Oversight Process, focuses on the areas of greatest safety significance and provides a more objective, risk-informed, transparent, predictable, and understandable approach to NRC's oversight role. Key features of the new program are a risk-informed regulatory framework, risk-informed inspections, a significance determination process to evaluate inspection findings, performance indicators, an objective and predictable assessment process, and more clearly defined actions the NRC will take for plants based on their performance. The NRC continues to use other supporting programs to assess potential plant performance issues. These programs include the Accident Sequence Precursor Program, the Program for Resolving Generic Issues, and programs for resolving allegations and petitions.

Also of interest to the NRC and its licensees are issues concerning utility restructuring that have arisen from economic deregulation of the U.S. electric power industry, which is proceeding on a State by State basis, and industry consolidation. The NRC staff recently performed a preliminary assessment of the effects of nuclear industry consolidation on the NRC and to determine whether the agency needs to change its regulations, policies, processes, guidance, or organizational structure to continue to meet its strategic public health and safety goals. The initial objective of this assessment was to identify potential effects that need to be considered further. The assessment determined that, among the issues important to safety, are grid reliability, the potential effect of cost-cutting measures taken by licensees to compete in a deregulated electric industry, and the potential benefit of large companies focused on nuclear power. The NRC has worked with the industry to address the reliability of the electric grid in supplying offsite power to nuclear power plants. Faced with the need to be competitive, licensees are examining the costs of current regulatory requirements and recommending alternative approaches for the NRC's consideration. For its part, the NRC has acted proactively to reduce unnecessary regulatory burden, for example, by improving its regulatory system to be more risk-informed. Thus far, it appears that industry restructuring is leading to significant consolidation of nuclear reactors under fewer, larger companies for whom nuclear power is a primary focus.

The increasing need for additional power in the U.S. and the improved economic and safety performance of nuclear power plants over the past decade have caused many licensees to consider renewing their licenses for an additional 20 years of operation, and to desire a timely review of their renewal applications by the NRC. In response, the NRC has made license renewal a high priority. Having gained efficiencies from its initial reviews of applications for the Calvert Cliffs Nuclear Power Plant, the Oconee Nuclear Station, and the Arkansas Nuclear One Plant, the NRC maintains an ambitious schedule to complete its reviews within 30 months of receiving an application if a hearing is held, and within 25 months if not. The agency has also recently completed an overhaul of its guidance documents for the license renewal process, an action that should improve the agency's effectiveness and efficiency in the years ahead. The NRC recently received license renewal applications involving a total of 10 units, which raises the total number of reactor units under various stages of review for license renewal to 14. On the basis of licensee-provided information, the agency expects licensees of an additional 24 reactor units to apply for license renewal over the next three years. If so, by the end of 2003, licensees of 44 of the 104 power reactors holding operating licenses will have applied for license renewal.

The NRC has also pursued extensive research on plant aging mechanisms to provide the technical bases to ensure that critical reactor components, safety systems, and structures remain reliable as reactors age. Toward that end, it uses the research results to assess the safety implications of age-related degradation during the period of the initial license, and to support safety decisions for license renewal.

The NRC has also been preparing for the potential licensing of new reactors. The agency has already issued certification rules for three new designs, the General Electric Advanced Boiling Water Reactor, the ASEA Brown Boveri/Combustion Engineering System 80+ (ABB-CE is now a part of BNFL), and Westinghouse AP600 (also part of BNFL). In addition, the NRC formed a group to prepare for and manage future reactor and site licensing applications. The group's work includes proposed rule changes to update the licensing framework, the pre-application review of the AP 1000, preparation for reviewing other new reactor designs, possible early site permits, and possible combined licenses. The NRC is also assessing the staff's technical, licensing, and inspection capabilities and identifying enhancements, if any, that would be necessary to ensure that the agency can effectively carry out its responsibilities associated with an early site permit application, a combined operating license application, and the construction of a new nuclear power plant.

The NRC has achieved other major accomplishments since writing the original U.S. National Report. These include issuing the Environmental Standard Review Plan; the final rule requiring assessment and management of risk in maintenance activities; the final rule on changes, tests, and experiments within licensee controls; the final rule recognizing that improved flow meter technology can reduce conservatism in evaluation models for the emergency core cooling system and thereby permit small power uprates across the industry; the final rule permitting the use of an alternative source term; and the final rule requiring consideration of potassium iodide as a supplemental protective action in emergency planning. The staff has also completed 57 reviews for plants requesting to increase the reactor power level ("power uprates"). As of August 2001, the U.S. has gained the power production equivalent of about two nuclear power plant units (of about 1000 MWe each) by implementing power uprates at existing plants.

INTRODUCTION

This section describes the purpose and structure of the report, the U.S. national policy towards nuclear activities, the main national nuclear programs, and the current nuclear safety initiatives. It then highlights major regulatory accomplishments since the original U.S. National Report was written. Finally, it references the list of nuclear installations in the U.S., which appears in Annex 1 to this report.

Purpose and Structure of this Report

This updated report is the submission by the United States of America to the second review meeting of the Contracting Parties to the Convention on Nuclear Safety for peer review. (This meeting is scheduled to be held at the International Atomic Energy Agency (IAEA) in Vienna, Austria, in April 2002.) The scope of the report considers only the safety of land-based civil nuclear power plants, consistent with the definition of nuclear installations provided in Article 2 and the scope of Article 3 of the Convention.

This report demonstrates how the U.S. Government meets the objectives described in Article 1 of the Convention, as follows:

(i) to achieve and maintain a high level of nuclear safety worldwide through the enhancement of national measures and international cooperation including, where appropriate, safety-related technical cooperation

(ii) to establish and maintain effective defenses in nuclear installations against potential radiological hazards in order to protect individuals, society, and the environment from harmful effects of ionizing radiation from such installations

(iii) to prevent accidents with radiological consequences, and to mitigate such consequences should they occur

Technical and regulatory experts from the U.S. Nuclear Regulatory Commission (which, in this report, is referred to as the NRC, Commission, agency, or staff) updated the original U.S. National Report, principally using NRC information that is publicly available. This updated report follows the format of the original U.S. Report on Nuclear Safety, submitted in 1998. Chapters are numbered according to the article of the Convention under consideration. Each chapter begins with the text of the article, followed by an overview of the material covered by the chapter, and a discussion of how the U.S. meets the obligations of the article. The report begins with Article 6, and continues through Article 19; annexes and references contain more detailed information. Articles 6 through 9 summarize the legislative and regulatory system governing the safety of nuclear installations and discuss the adequacy and effectiveness of that system. Articles10-16 address general safety considerations and summarize major safety-related features. Articles 17-19 address the safety of installations. This report does not include chapters for Articles 1 through 5. In accordance with Article 1, the report illustrates how the U.S. Government meets the objectives of the Convention. It discusses the safety of nuclear installations according to their definition in Article 2 and the scope of Article 3. It addresses implementing measures (such as national laws, legislation, regulations, and administrative means) according to Article 4. Submission of the report fulfills the obligation of Article 5 on reporting.

The U.S. National Policy Towards Nuclear Activities

The NRC was created by enactment in the Congress of the Energy Reorganization Act of 1974. As such, the NRC is an independent agency of the Federal Government. The mission of the NRC is to ensure that civilian uses of nuclear materials in the United States — in the operation of nuclear power plants and fuel cycle plants, and in medical, industrial, and research applications — are carried out with proper regard and provision for the protection of public health and safety, the environment, and national security. The agency also has a role in combating the proliferation of nuclear materials worldwide. The safety and security responsibilities of the NRC stem from the Atomic Energy Act of 1954. The NRC accomplishes its mission by licensing and overseeing nuclear reactor operations and other activities that apply to the possession of nuclear materials and wastes, safeguarding nuclear materials and facilities from theft and radiological sabotage, issuing rules and standards, inspecting nuclear facilities, and enforcing regulations.

United States laws require that the nuclear regulatory program be carried out with full participation of the public. Except for certain vendor-proprietary business material, facility safeguards information, and information supplied by foreign countries that is deemed to be sensitive, the documentation produced by the NRC is available in the agency's Public Document Room in Rockville, Maryland, and on its website at *http://www.nrc.gov*. As a result, nuclear activities and the national policy toward them are open to everyone.

National Nuclear Programs

The NRC uses a number of programs and processes to carry out its mission and meet the obligations of the Convention of Nuclear Safety. These are a well-established licensing process; the newly revised process for reactor oversight; programs for responding to and evaluating operational events, including the Accident Sequence Precursor Program and the Program for Resolving Generic Safety Issues; programs for rulemaking; for decommissioning; for regulatory research; for public participation; and for handling petitions, allegations, and differing professional views and opinions.

The NRC continues to refine its Reactor Oversight Program to make it more risk-informed, objective, predictable, understandable, and focused on areas of greatest risk significance. Key features of the program are a risk-informed regulatory framework and inspections, a process to determine the significance of inspection findings, the use of performance indicators, an objective and predictable process to assess plant performance, and more clearly defined actions that the NRC will take for plants on the basis of their performance.

The NRC uses the Accident Sequence Precursor Program to evaluate the conditional probability (i.e., probability, given an initiating condition) of core damage from plant events and unavailability of equipment. The program provides a structured, systematic way of quantitatively evaluating the safety significance of nuclear plant operating experience. The program's key features are identifying and ranking the risk significance of operating reactor events, determining their generic implications, characterizing risk insights, and routing evaluations for feedback to plant operators to promote learning from experience.

The NRC uses the Program for Resolving Generic Issues to resolve issues concerning nuclear power plants that arise from such sources as safety-related research, risk-assessment analyses, and public and industry concerns.

For more details on all of these programs see Article 6.

Main Nuclear Safety Initiatives

The NRC and its licensees are implementing the following important initiatives in 2001:

- development of risk-informed regulatory programs
- implementation of revised reactor oversight process
- assessment of the implications of industry restructuring and consolidation
- enhancement of license renewal regulatory infrastructure and pursuit of aging research
- preparation for potential licensing of new reactors

Development of Risk-informed Regulatory Programs

Benefitting from over 40 years of operational experience and steadily improving plant performance by licensees, the NRC has adopted a more risk-informed regulatory approach. The current regulatory framework is based largely, but not entirely, on a "deterministic" approach that employs safety margins, operating experience, accident analyses, and qualitative assessments of risk, and relies on a defense-in-depth philosophy. (A deterministic approach refers to an approach that specifies certain design and operational conditions and applies bounding criteria to demonstrate acceptable plant performance.) In 1995, the NRC adopted a policy that promotes increasing the use of probabilistic risk analysis in all regulatory matters to the extent supported by the state-of-the-art to complement the deterministic approach. This complementary aspect is why the NRC refers to its actions as being "risk-informed."

Using risk insights, the NRC has modified its oversight process and its requirements for maintenance in Title 10, Section 50.65 of the _U.S. Code of Federal Regulations_ (10 CFR 50.65). The NRC is considering further revisions to its reactor regulations (10 CFR Part 50) to focus requirements on programs and activities that are most risk significant. However, these revisions would provide alternatives that are strictly voluntary to current requirements. The agency is also considering changes to 10 CFR Part 50 that could lead to incorporating a new set of design-basis accidents, revising specific requirements to reflect risk-informed considerations, or deleting certain regulations. The move toward risk-informing the current regulations and processes marks perhaps the most significant changes taking place at the NRC.

Current activities in risk-informed regulation include proposals to update certain important technical regulations. These activities include recommending risk-informed changes to 10 CFR 50.44,"Standards for Combustible Gas Control System in Light-Water-Cooled Power Reactors" and 10 CFR 50.46 (Emergency Core Cooling System Acceptance Criteria).

In addition, the staff continues to update Regulatory Guide 1.174, "An Approach for Using Probabilistic Risk Assessment in Risk-Informed Decisions on Plant-Specific Changes to the Licensing Basis." This guide describes an acceptable method for licensees and NRC staff to use when applying risk information to assess the nature and effect of the proposed licensing basis.

The NRC has also approved pilot applications in graded quality assurance, inservice testing of pumps and valves, and inservice inspection of important reactor plant piping. In each of these applications, risk information provides a strong basis for a graded treatment of the regulated activities at certain licensee facilities. This information allows both the NRC and licensees to focus resources on equipment and activities that have the greatest risk significance.

Article 10 explains the risk-informed regulatory approach in more detail.

Implementation of Revised Reactor Oversight Process

Recently, the NRC phased out its previous reactor assessment process and replaced it with a new integrated inspection, assessment and enforcement process known as the Revised Reactor Oversight Process. The goal of the Revised Reactor Oversight Process is to increase the predictability, transparency, consistency, and objectivity of the NRC's assessments of reactor performance. The NRC accomplishes this goal by considering certain quantitative reactor performance indicators and by using risk assessment to determine the safety significance of inspection findings. The agency restructured the inspection process to include a baseline inspection of every plant and more extensive inspections when called for by either performance indicators or inspection findings. The NRC continues to use supporting programs to assess potential plant performance issues. These programs include the Accident Sequence Precursor Program, the Program for Resolving Generic Issues, and programs for resolving allegations, and handling public petitions submitted under 10 CFR 2.206. Article 6 explains the programs in detail.

Assessment of the Implications of Industry Restructuring and Consolidation

Although no proposed Federal legislation on the economic deregulation of power plants has yet become law, many States have already moved to deregulate the retail electricity generation market. This evolving environment presents several challenges to the NRC, and the agency is currently monitoring and responding to these challenges.

The NRC has been dealing with utility restructuring and financial assurance issues that have arisen as a result of economic deregulation and industry consolidation. Among the potential safety issues are grid reliability, the effect of cost cutting measures taken by licensees to compete in a deregulated electric industry, and the benefit of large companies focused on nuclear power.

Deregulation has created an increasing need to ensure electrical grid reliability, especially as Independent System Operators are established. The NRC has worked with the industry to address the effect of utility deregulation on the reliability of the electric grid in supplying offsite power to nuclear power plants. The NRC is cooperating with the Federal Energy Regulatory Commission, as well as other agencies, on this reliability issue.

Faced with increased cost pressures from rate deregulation and the need to be competitive, licensees are evaluating the costs of regulatory requirements and are recommending alternative approaches for NRC's consideration. For its part, the NRC has acted proactively to reduce unnecessary regulatory burden, for example, by improving its regulatory system to be more risk-informed.

As the industry has restructured, the NRC has received requests to approve different types of operating and ownership arrangements. Most of these requests entail transferring the NRC license. Because the NRC reviews the new business and ownership arrangements, the agency has issued guidance on conducting reviews and provided uniform rules of practice for prompt handling of hearing requests for license transfer applications. The NRC has also approved over ten applications since 1999 for the transfers of nuclear plant licenses, including the sale of the entire plant. In one such transaction, AmerGen, a consortium of Philadelphia Electric Company Energy, Inc. (PECO Energy), and British Energy, PLC, purchased the Three Mile Island Unit 1 reactor, and is selling power from the plant at market-based prices. In every instance, license transfers have led to consolidation of nuclear reactors in larger companies for which nuclear power is a primary focus.

The NRC has also expanded allowable methods for providing decommissioning funding assurance. Towards that end, the agency published the final rule on decommissioning funding assurance in September 1998. Developed to account for the deregulation of the electric utility industry, the rule allows licensees to use external sinking fund methods of financial assurance for decommissioning exclusively if they continue to be rate regulated. The rule also allows additional financial assurance mechanisms for licensees to use at their discretion or if they are no longer rate regulated. In addition, the rule allows licensees to take credit on earnings for prepaid decommissioning trust funds, and requires them to periodically report to the NRC on the status of these trust funds.

Article 11 explains restructuring and financial assurance issues in more detail.

Enhancement of License Renewal Regulatory Infrastructure and Pursuit of Aging Research

The increasing need for additional power in the U.S. and the improved economic and safety performance at nuclear power plants over the last decade have caused many licensees to consider license renewal for an additional 20 years of operation and to desire a timely review of applications by the NRC. In response, NRC has made license renewal a high priority. Having gained efficiencies from its initial reviews of applications for Calvert Cliffs Units 1 and 2 and the Oconee Units 1, 2, and 3, the NRC maintains an ambitious schedule to complete reviews within 30 months of receiving a license renewal application if a hearing is held, and within 25 months if not. The NRC recently issued the renewed license for Arkansas Nuclear One, Unit 1. The NRC is now reviewing license renewal applications for Hatch, Turkey Point, Surry, North Anna, McGuire, Catawba, and Peach Bottom, a total of 14 Units. The NRC expects to have more than 10 license renewal applications in various stages of review in the near future. Almost all current nuclear industry licensees have expressed an interest in renewing their licenses in the future. The NRC has recently completed the overhaul of all of its guidance documents for the license renewal process. This action should significantly improve the effectiveness and efficiency of the agency in the years ahead.

The NRC has conducted research to provide the technical bases to ensure that critical reactor components, safety systems, and structures remain reliable as reactors age. Towards that end, the agency uses research results to assess safety implications of age-related degradation during the period of the initial license, and to support safety decisions for license renewal.

Article 14 explains license renewal in more detail.

Preparation for Potential Licensing of New Reactors

In 1989, the NRC revised its regulations to provide a more stable and predictable process for licensing nuclear power plants. This process includes the use of early site permits, standard design certifications, and combined licenses. Under this process, the NRC has already issued certification rules for three new designs, namely the General Electric Advanced Boiling Water Reactor, the ASEA Brown Bovari/Combustion Engineering (now Westinghouse) System 80+, and the Westinghouse AP600. The NRC is now expecting new applications for early site permits, design certifications, and combined licenses. Next year, the NRC expects an application for standard design certification of the Westinghouse AP1000 design. In addition, several organizations have contacted the NRC to discuss building and licensing new nuclear power plants in the U.S. These include Exelon Generation, which has requested a pre-application review of the pebble bed modular reactor, and several other organizations, which may apply for early site permits in the next few years.

To ensure that the agency can effectively carry out its responsibilities associated with an early site permit application, a license application, and the construction of a new nuclear power plant, the NRC has formed an organization to: (1) manage near term future licensing activities (regulatory infrastructure changes, the AP1000 pre-application review, the pre-application review for the pebble bed modular reactor); (2) work with stakeholders regarding new reactor licensing activities; and (3) assess the NRC's readiness to perform new reactor licensing reviews. The staff has almost completed the readiness assessment. This assessment will address: (1) licensing scenarios for future application reviews, the durations of the reviews, and resource estimates to complete the reviews; (2) critical skills that must be available within the agency or that can be accessed through contractual agreements to perform these reviews; and (3) regulatory infrastructure changes necessary to support future licensing reviews. The establishment of this organization is a two-phase process. Initially, staff members, some of whom have experience with standard and advanced reactor reviews and environmental reviews, were temporarily assigned. By the end of 2001, the NRC will assign a permanent staff that will continue these initial efforts and carry out the tasks established as a result of the readiness assessment.

Other Major Regulatory Accomplishments

In April 1999, the NRC revised 10 CFR Part 55, "Operators' Licenses," to allow nuclear power reactor licensees to prepare the written examinations and operating tests that the NRC uses to evaluate the competence of applicants for operators' licenses. The NRC will review licensee-prepared examinations, and continue to administer all operating tests and make the final licensing decisions. This rule change will maintain nuclear reactor safety, while increasing staff effectiveness and efficiency. (See Article 11 for more details on the licensing of operators.)

In July 1999, the NRC revised 10 CFR 50.65, "Requirements for Monitoring the Effectiveness of Maintenance at Nuclear Power Plants," ("the Maintenance Rule") to require that licensees, before performing maintenance, assess and manage the increases in risk that may result from the maintenance activities. Before the revision, the Maintenance Rule recommended, but did not require, performing safety assessments. Article 19 provides more details on the Maintenance Rule.

In October 1999, the NRC revised the rule governing design changes, 10 CFR 50.59, "Changes, Tests, and Experiments." Under this rule, licensees can make certain changes to their facilities without prior NRC approval. The revision to the rule is meant to clarify NRC requirements, and to allow changes that will minimally affect the facility licensing basis. Article 14 provides more details on 10 CFR 50.59.

In March 2000, the NRC issued the "Environmental Standard Review Plan," which is segregated into two documents: a "greenfield" (new site) review plan and a license renewal review plan. Environmental reviews complement the safety reviews and stem from the National Environmental Policy Act of 1969. Article 17 explains environmental reviews.

In May 2000, the Commission approved a final rule amending 10 CFR Part 50, Appendix K, "ECCS Evaluation Models" to facilitate small, but cost-beneficial, power uprates for some commercial nuclear power plants through the use of improved flow meter technology. The rule amendment changes a provision requiring emergency core cooling system (ECCS) performance analyses to assume the reactor to be operating 2 percent above licensed power. The amendment allows licensees to adopt an alternative power level to the value stated in the rule if the alternative is sufficiently justified. Article 6 explains the rule in more detail.

In December 2000, the NRC published the final rule, 10 CFR 50.67, "Accident Source Term," permitting licensees of operating reactors to use an alternative source term in lieu of the traditional source term. (The source term refers to the release of fission products from the reactor core into the containment during an accident). Article 17 explains the rule and the source term in more detail.

In January 2001, the NRC published the final rule, 10 CFR 50.47, "Consideration of Potassium Iodide in Emergency Plans," requiring the consideration of potassium iodide prophylaxis as a supplemental protective action in emergency planning. Article 16 explains the rule in more detail.

Nuclear Installations in the U.S.

Annex 1 contains a list of the nuclear installations in the U.S.

ARTICLE 6. EXISTING NUCLEAR INSTALLATIONS

Each Contracting Party shall take the appropriate steps to ensure that the safety of nuclear installations existing at the time the convention enters into force for that Contracting Party is reviewed as soon as possible. When necessary in the context of this convention, the Contracting Party shall ensure that all reasonable practicable improvements are made as a matter of urgency to upgrade the safety of the nuclear installation. If such upgrading cannot be achieved, plans should be implemented to shut down the nuclear installation as soon as practically possible. The timing of the shutdown may take into account the whole energy context and possible alternatives as well as the social, environmental, and economic impact.

This section explains how the U.S. ensures the safety of nuclear installations in accordance with the obligations in Article 6. First, it summarizes the characteristics of the nuclear industry in the U.S. Then, it explains reactor licensing. Next, it discusses the major oversight process in the U.S. -- the Reactor Oversight Process -- and supporting programs: the Accident Sequence Precursor Program and the Program for Resolving Generic Issues. Then, it discusses programs for rulemaking, decommissioning, and research; and programs for public participation, handling petitions, resolving allegations, and settling differing professional views and opinions. The "Experience and Examples" subsections cover nuclear installations for which the NRC's assessments showed that corrective actions were necessary. The NRC posts the major results of assessments on its Website at *http://www.nrc.gov*

The safety performance of the U.S. nuclear power industry has improved substantially over the past ten years, and nuclear reactors, collectively, are operating above acceptable safety levels consistent with the agency's Safety Goal Policy. As a result, the NRC has established maintaining this level of safety as its most important performance goal. If the agency identifies the need for substantial safety improvements, it will only impose additional requirements consistent with the Commission's Backfit Rule (10 CFR 50.109).

6.1 Nuclear Installations in the U.S.

As of December 1999, 104 commercial nuclear power reactors are licensed to operate in 31 States. Although similar in many ways, each reactor design is unique. The U.S. nuclear power industry is composed of 4 reactor vendors, 30[1] licensees, 80 different designs, and 65 sites. The 104 reactors licensed to operate during 2000 have accumulated 2150 reactor-years of experience; permanently shutdown reactors have accumulated an additional 357 reactor-years. For a list of nuclear installations in the U.S., see Annex 1.

---

[1]The number of licensees has fallen from 45 in 1996 to 30 on September 1, 2001, as consolidation has proceeded. The number of licensees is expected to be further reduced in the years ahead.

6.2 Regulatory Processes and Programs

This section discusses the processes and programs the NRC uses to ensure that plant safety is maintained. These are a well established licensing process; a newly revised process for reactor oversight; programs for responding to and evaluating operational events, including the Accident Sequence Precursor Program and the Program for Resolving Generic Safety Issues; and programs for rulemaking, for decommissioning, for regulatory research, for public participation, and for handling petitions, allegation, and differing professional views and opinions.

6.2.1 Reactor Licensing

To construct and operate a nuclear reactor, an entity must submit an application to the NRC for safety and environmental review. The public also has opportunities to participate through a hearing process. The detailed two-step review performed by the NRC to issue a construction permit and then an operating license under 10 CFR Part 50, "Domestic Licensing of Production and Utilization Facilities," is described under Article 18. Although the NRC has not received any applications for new reactor licensing since 1976, the agency has issued new regulations for early site permits, standard design certification and combined licenses for nuclear power plants (10 CFR Part 52) to streamline the licensing process. In addition, the licensing process provides for the review and approval of changes after initial licensing. These provisions address the issuing of amendments to the operating license to support plant changes, license renewal, changes of ownership and license transfer, exemptions and relief from NRC regulations, and increasing the reactor power level ("power uprates"). These provisions are discussed in other articles, except for power uprates, which are discussed below.

Power uprates can be classified in three categories: (1) measurement uncertainty recapture power uprates, (2) stretch power uprates, and (3) extended power uprates. On the order of 1.5 percent, measurement uncertainty recapture power uprates are achieved by implementing enhanced techniques for calculating reactor power. These techniques involve using state-of-the-art feedwater flow measurement devices that reduce the degree of uncertainty associated with feedwater flow measurement and in turn provide for a more accurate calculation of power. The recent rulemaking to 10 CFR Part 50, Appendix K, "ECCS (Emergency Core Cooling System) Evaluation Models," which allowed licensees to use a power uncertainty less than 2 percent in loss-of-coolant accident analyses, facilitated reviews of these uprates. Stretch power uprates are typically on the order of 7 percent and usually involve changes to instrumentation setpoints. Stretch power uprates do not generally involve major plant modifications, particularly for boiling-water reactor plants. In some limited cases in which plant equipment was operated near capacity before the power uprate, more substantial changes may be required. Extended power uprates are usually greater than stretch power uprates, and requests are expected for increases as high as 20 percent. Extended power uprates usually require significant modifications to major balance-of-plant equipment such as the high pressure turbines, condensate pumps and motors, main generators, and/or transformers.

Licensees have been applying for and implementing power uprates since the 1970s as a way to increase the power output of their plants. The staff has been reviewing applications for power uprates since then and has completed 57 such reviews. As of August 2001, the staff had approved measurement uncertainty recapture power uprates for 4 units, stretch power uprates for 50 units, and extended power uprates for 3 units. As of August 2001, the U.S. has gained the power production equivalent of about two nuclear power plant units (of about 1000 MWe each) by implementing power uprates at existing plants.

As of August 2001, the staff had 17 applications for power uprates under review. Of these, nine were for measurement uncertainty recapture power uprates, and six were for extended power uprates greater than or equal to 15 percent. The remaining two applications included one for 4.5 percent and one for 7.5 percent.

On April 2, 2001, the staff issued NRC Regulatory Issue Summary 2001-08, "Operating Reactor Licensing Action Estimates." In this document, the staff requested, on a voluntary basis, information related to future submittals of licensing actions for fiscal years 2001 and 2002. In addition, the staff surveyed all licensees in June 2001 to obtain information regarding the industry's future plans related to power uprate applications. The survey targeted projections for the size of power uprates and schedule of submittals over the next 5 years. The results of this survey indicate that licensees plan to submit 46 power uprate applications in the next 5 years. Of these, 15 are expected for extended power uprates, 3 are expected for stretch power uprates, and 27 are expected for measurement uncertainty recapture power uprates. (One licensee did not report a size for the expected uprate.) The sizes reported for the stretch and extended power uprates may also include measurement uncertainty recapture. On the basis of the information provided, the NRC expects the planned power uprates to result in an increase of about 4870 MWt or about 1600 MWe. The staff will use the information provided in response to the regulatory issue summary and survey to plan and allot resources for power uprate reviews and to ensure the staff's readiness and availability to perform the technical reviews of the applications when they arrive.

6.2.2 Reactor Oversight Process

The NRC continuously oversees nuclear power plants to verify that they are being operated in accordance with the agency's rules and regulations. The NRC has full authority to take whatever action is necessary to protect public health and safety, and may demand immediate licensee actions, up to and including a plant shutdown.

This section explains the reactor oversight process. It covers the goals, regulatory framework, inspections and performance indicators, significance determination process, enforcement, how the current oversight process differs from the previous process, and experience and examples.

6.2.2.1 Goals

The NRC recently revised its Reactor Oversight Process. The revised process is intended to fulfill the following four goals established by the Commission:

(1) Maintain safety by establishing a regulatory oversight framework that provides assurance that licensees continue to operate plants safely.

(2) Enhance public confidence in the NRC's regulatory process by increasing the predictability, consistency, objectivity, and transparency of the oversight process.

(3) Improve the effectiveness, efficiency, and realism of the oversight process by focusing both the agency's and licensees' resources on issues that are most significant to safety.

(4) Reduce unnecessary regulatory burden.

6.2.2.2 Regulatory Framework

The regulatory framework for reactor oversight is shown in Figure 1. A risk-informed, tiered approach for ensuring plant safety, this framework has three key strategic performance areas: reactor safety, radiation safety, and safeguards. Within each strategic performance area are "cornerstones" that apply to the essential safety aspects of plant operation. Satisfactory licensee performance in the cornerstones provides reasonable assurance that plants are being safely operated, and that the NRC's safety mission is being accomplished. Within this framework, the NRC's operating reactor oversight process provides a way of collecting information about licensee performance, assessing the information for its safety significance, and providing for appropriate licensee and NRC responses.

To measure plant performance, the oversight process focuses on the following seven specific "cornerstones" that support the safety of plant operations in the three key strategic areas:

- Initiating Events - This cornerstone focuses on operations and events at a nuclear plant that could lead to a possible accident if plant safety systems did not intervene. These events include equipment failures leading to a plant shutdown, shutdowns with unexpected complications, or large changes in the plant's power output.

- Mitigating Systems - This cornerstone focuses on the function of safety systems that are designed to prevent an accident or reduce the consequences of a possible accident.

- Barrier Integrity - This cornerstone focuses on the licensee's effectiveness in maintaining the three physical barriers – fuel rods, reactor vessel and associated piping, and containment.

- Emergency Preparedness - This cornerstone focuses on the effectiveness of the plant's staff in carrying out its emergency plans.

- Public Radiation Safety - This cornerstone focuses on the effectiveness of the licensee's programs to meet applicable Federal limits involving the exposure, or potential exposure, of members of the public to radiation and ensure that the effluent releases from the plant are as low as reasonable achievable.

- Occupational Radiation Safety - This cornerstone focuses on the effectiveness of the licensee's program(s) to maintain the worker dose within the regulatory limits and provide occupational exposures that are as low as reasonably achievable.

- Physical Protection - This cornerstone focuses on the effectiveness of the security, safeguards, and fitness-for-duty programs.

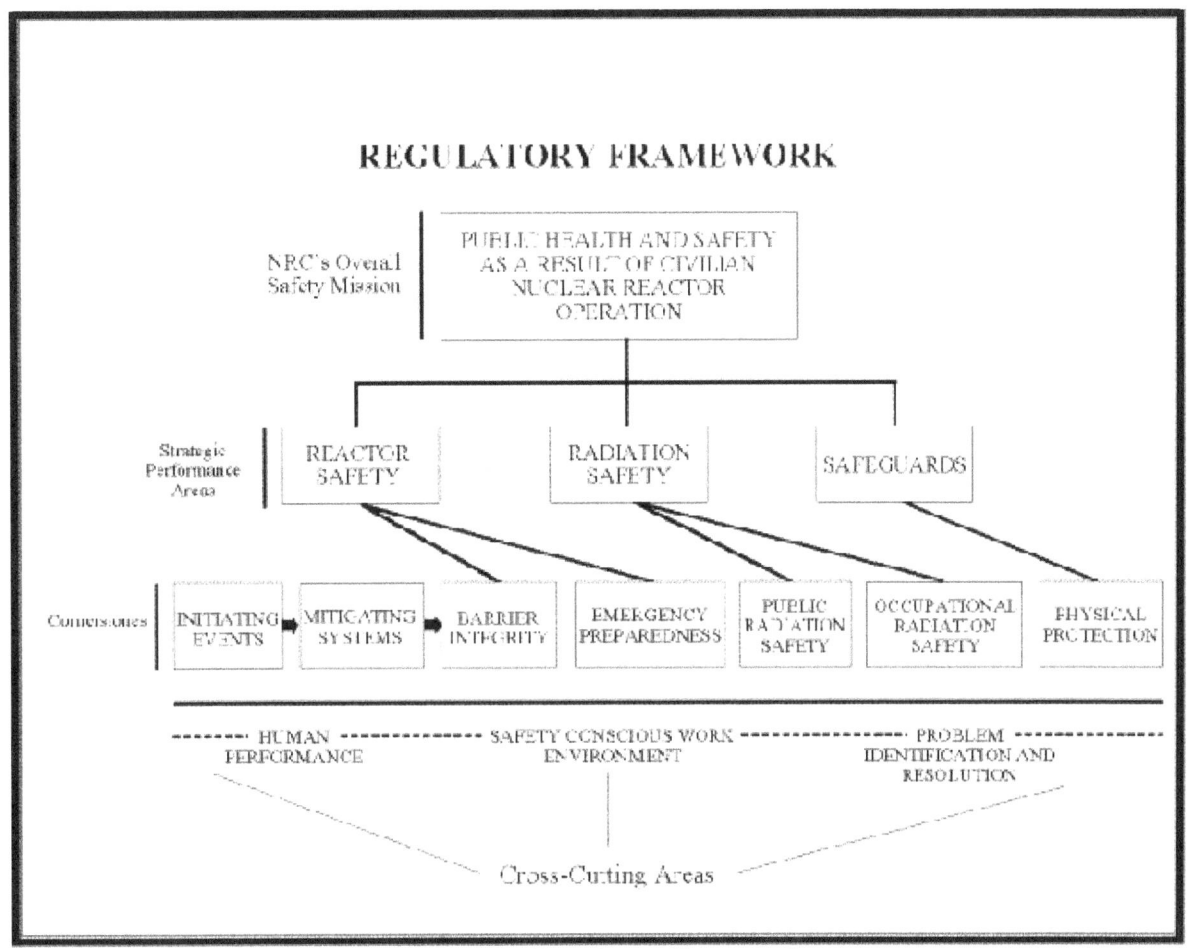

Figure 1: Regulatory Framework

In addition to the cornerstones, the reactor oversight framework recognizes three "cross-cutting" elements, so named because they affect, and are therefore part, of each of the cornerstones:

- human performance
- safety-conscious work environment
- problem identification and resolution

The assessment of these cross-cutting elements is important to the oversight process.

Within each of the "cornerstone" areas, the NRC designed performance indicators and inspections to closely focus on plant activities most affecting safety and overall risk. (Table 1 shows the various cornerstones and performance indicators.)

Table 1: Performance Indicators by Cornerstone

Safety Cornerstone	Performance Indicator
Initiating Events	Unplanned reactor shutdowns (automatic and manual)
	Loss of normal reactor cooling system following unplanned shutdown
	Unplanned events that results in significant changes in reactor power
Mitigating Systems	Safety system not available • specific emergency core cooling systems • emergency electric power systems
	Safety system failures
Integrity of Barriers to Release of Radioactivity	Fuel Cladding (measured by radioactivity in the reactor cooling system)
	Reactor cooling system leak rate
Emergency Preparedness	Drill and exercise performance
	Emergency response organization drill participation
	Alert and Notification System Reliability
Public Radiation Safety	Effluent release requiring reporting under NRC regulations and license conditions
Occupational Radiation Safety	Compliance with requirements for the control of access to areas of the plant with dose rates greater than one rem/hr (10 mSv/hr.) Unintended radiation exposures to workers greater than a specified fraction of the dose limits in 20. 1201, 20.1207, and 20.1208[2]
Physical Protection	Security system equipment availability
	Personnel screening program performance
	Employee fitness-for-duty program effectiveness

An example of a performance indicator in the Emergency Preparedness cornerstone is "Alert and Notification System Reliability." This performance indicator monitors the reliability of the alert and notification system, which has been identified as the most-risk significant equipment system for emergency preparedness programs of nuclear plants. This indicator measures the percentage of system sirens that are capable of performing their function as measured by periodic testing, (e.g., silent, growl, siren sound test) in the previous 12 months.

---

[2] 2% of the stochastic limit in 10 CFR 20.1201, for total effective dose equivalent, 10% of the non-stochastic limit in 10 CFR 20.120, for individual organ doses, 20% of the limits in 10 CFR

20.1207 and 10 CFR 20.208, for minors and declared pregnant workers respectively.

6.2.2.3 Inspections and Performance Indicators

Although performance indicators can give insights into plant performance for selected areas, the NRC's inspection program provides a greater depth and breadth of information. The NRC designed the baseline inspection program to verify the accuracy of performance indicator information, and to assess performance that is not directly measured by the performance indicator data.

The NRC uses a "risk-informed" approach to select areas to inspect within each cornerstone. In so doing, the agency selects inspection areas because of their importance to potential risk, past operational experience, and regulatory requirements.

Common to all nuclear plants and targeted around the "cornerstone" areas, baseline inspections focus on activities and systems that are "risk significant," (that is, activities and systems that could trigger an accident, mitigate the effects of an accident, or increase the consequences of a possible accident). Baseline inspections are of three kinds -- inspection of areas that are not covered or not fully covered by performance indicators, inspections to verify the accuracy of a licensee's reports on performance indicators, and inspections that consist of a thorough review of the licensee's effectiveness in finding and resolving problems on its own. In baseline inspections, inspectors, also examine the "cross-cutting issues" of human performance, the "safety-conscious work environment," and the licensee's ability to identify and correct problems. The NRC performs inspections that go beyond baseline inspections at plants where performance falls below established performance indicator thresholds or when inspections reveal significant findings, or to respond to a specific event or problem that may arise at a plant.

The NRC evaluates the performance indicator data, and integrates it with the findings of the agency's inspection program. Each of the performance indicators has thresholds for measuring acceptable performance. These thresholds indicate risk according to established safety margins, as indicated by a color coding system.

A "green" coding indicates performance within an expected performance range in which the related cornerstone objectives are met; "white" indicates performance outside an expected range of nominal licensee performance, but related cornerstone objectives are still being met; "yellow" indicates that related cornerstone objectives are being met, but with a minimal reduction in safety margin; and "red" indicates a significant reduction in the safety margin in the area measured by that performance indicator. Each licensee reports the performance indicators to the NRC on a quarterly basis.

The NRC measures each performance indicator against established thresholds that are related to their effects on safety, and uses a similar color coding system to classify performance associated with that indicator. For example, thresholds for "Alert and Notification System Reliability" are less than 94% system availability for white coded findings, and less than 90% system availability for yellow coded findings, (availability meaning operational on demand).

6.2.2.4 Significance Determination Process

The NRC developed a process known as the "Significance Determination Process" to determine the safety significance of inspection findings. The staff uses this process to screen for inspection findings that do not result in a notable increase in risk and thus need not be further analyzed or require a regulatory response from the NRC ("green" findings). The staff then subjects the remaining inspection findings -- which may affect plant risk -- to a more thorough risk assessment, according to the next phase of the process. NRC risk experts from the cognizant regional office and headquarters may participate, and the licensee's plant staff may have to conduct further reviews. The final outcome of the review -- a finding of green, white, yellow, or red -- determines a further NRC response. Table 2 summarizes the NRC's responses.

Each calendar quarter, resident inspectors and the inspection staff in each regional office review the performance of all nuclear power plants in the given region, as measured by the performance indicators and inspection findings. Every 6 months, the NRC staff expands this review to plan for inspections for the following 12-month period.

During the final quarterly review for each year, the NRC assesses plant performance over the previous 12 months in more detail and prepares a performance report, and the inspection plan for the following year. The NRC posts these annual performance reports on its Web site, and holds public meetings with licensees to discuss the previous year's performance at each plant.

In addition, the NRC's senior management reviews the adequacy of the agency's planned actions for plants with significant performance problems. The managers also take a wider view of both the overall industry performance and the performance of the agency's regulatory programs. In addition, they discuss the performance of plants requiring heightened agency scrutiny during a public meeting with the NRC Commissioners at the agency's Headquarters in Rockville, Maryland.

In its quarterly reviews of plant performance, the NRC determines what additional action, if any, it will take if there are signs of declining performance. This approach to regulatory oversight is intended to be more predictable than previous practices by linking regulatory actions to performance criteria. The new process has four levels of regulatory response, according to which NRC regulatory review increases as plant performance declines. The cognizant regional office manages the first two levels of heightened regulatory review; the next two levels call for an agency response involving attention from senior management in both Headquarters and Regional offices. The NRC's actions for performance below "green" coding may include meetings with the licensee, additional inspections, and required reviews of responses by the licensee. Further declines in performance would warrant stronger action by the NRC, possibly including an Order that could modify or even suspend the licensee's operating license.

For results of assessments of plant performance, see *http://www.nrc.gov.*

Table 2: NRC Response Plan or "Action Matrix"

Assessment of Plant Performance (In order of increasing safety significance)	NRC Response
I. All performance indicators and cornerstone inspection findings GREEN • Cornerstone objectives fully met	• Routine inspector and staff interaction • Baseline inspection program • Annual assessment public meeting
II. No more than two WHITE inputs in different cornerstones	**Response at Regional level** • Staff to hold public meeting with licensee management • Licensee corrective action to address WHITE inputs • NRC inspection followup on WHITE inputs and corrective action
III. One degraded cornerstone (two WHITE inputs or one YELLOW input or three WHITE inputs in any strategic area • Cornerstone objectives met with minimal reduction in safety margin	**Response at Regional level** • Senior regional management to hold public meeting with licensee management • Licensee to conduct self-assessment with NRC oversight • Additional inspections focused on cause of degraded performance
IV. Repetitive degraded cornerstone, multiple degraded cornerstones, or multiple YELLOW inputs, or one RED input • Cornerstone objectives met with long-standing issues or significant reduction in safety margin	**Response at Agency level** • Executive Director for Operations to hold public meeting with senior licensee management • Licensee develops performance improvement plan with NRC oversight • NRC team inspection focused on cause of degraded performance • Consideration of Demand for Information, Confirmatory Action Letter, or Order
V. Unacceptable Performance • Unacceptable reduction in safety margin	**Response at Agency level** • Plant not permitted to operate • Commission meeting with senior licensee management • Order to modify, suspend, or revoke license

6.2.2.5 Enforcement of NRC Requirements

The NRC evaluates each violation of NRC requirements that it finds during inspections to determine the effect on plant safety and risk. If the evaluation reveals that the violation is of very low safety significance, the NRC discusses the violation in the inspection report and does not take any formal enforcement action. The licensee is expected to deal with the violation through its corrective action program, correcting the violation and taking steps to prevent a recurrence. The NRC may also review the issue during future inspections to determine the effectiveness of the licensee's corrective action.

If the NRC's risk evaluation reveals that the violation has higher safety significance, the agency will issue a Notice of Violation. The NRC may also issue a Notice of Violation if the licensee fails to correct a violation of low safety significance in a reasonable period of time, or if the agency finds that a violation is willful. The Notice of Violation requires the licensee to formally respond to the NRC about its actions to correct the violation, and to identify the steps it will take to prevent the violation from occurring again. The agency then reviews the licensee's actions in a later inspection. Normally, these violations will not be the subject of a monetary civil penalty (fine). However, there may be violations that warrant a fine because of their unusual significance, such as an accidental criticality. These violations are likely to be uncommon.

In addition, the following violations call for the traditional enforcement approach, including the use of severity levels to reflect the significance (versus the color coding according to the significance determination process) and possible fines:

- Willful violations, including discrimination against workers for raising safety issues.

- Actions that may adversely affect the NRC's ability to monitor licensee's activities, including failures to report required information, to obtain NRC approval for certain plant changes, to maintain accurate records, or to give the NRC complete and accurate information.

- Incidents with actual safety consequences, including radiation exposures above NRC limits, doses above NRC limits resulting from releases of radioactive material, or failure to notify government agencies when emergency response is required.

6.2.2.6 How This New Oversight Process Differs from the Previous Process

The previous oversight process evolved over a period of time when the nuclear power industry was less mature and there was much less operational experience on which to base rules and regulations. Conservative judgments governed the rules and regulations and the oversight process tended to be reactive and prescriptive. The process was characterized by the NRC closely observing plant performance for the licensee's adherence to the regulations, and responding to operational problems as they occurred. Today, however, the nuclear power industry has the benefit of four decades of operational experience and steadily improving plant performance, particularly over the last decade or so.

The new process focuses more of the agency's resources on the relatively small number of plants with performance problems. The baseline inspection program is considered the minimum inspection effort needed to ensure that plants meet the "safety cornerstone" objectives. The NRC will more thoroughly inspect plants that do not meet the objectives, focusing on areas of declining performance. If the NRC believes that there are specific operational problems or events that require greater scrutiny, then it performs inspections beyond the baseline program. The agency may also perform additional inspections of generic problems, problems that affect some or all nuclear plants.

In the new oversight process, the NRC maintains the same enforcement tools that it used in the past for dealing with declining plant performance and violations. The NRC uses these tools, however, more predictably and in a manner that is commensurate with the decreased safety performance. Under the new process, the NRC has a system of specified agency actions if licensee performance declines. In the past, the agency tended to use fines as a prime indicator of agency concern and as a motivator to effect licensee corrective actions. The NRC now uses enforcement as one part of an overall regulatory process. It may use other regulatory tools to respond to performance issues, such as increased inspections, management meetings, demands for information, confirmatory action letters, or orders. The agency reserves fines for those violations that involve willfulness, or can affect the agency's ability to perform its regulatory function, or result in an actual adverse consequence.

The new assessment process is substantially different from the previous one in that it makes greater use of objective performance indicators. Together, the indicators and inspection findings supply the information needed for the quarterly reviews of plant performance. The most important differences between the new reactor oversight process and the previous assessment process are: (1) the expanded use of performance indicators in addition to the inspection findings, (2) the use of a risk-informed process to determine the significance of inspection findings, and (3) the use of inspection results and the performance indicators to determine the level of regulatory response in the applicable safety cornerstone area. The new process provides a more objective, risk-informed, predictable and understandable approach to NRC's regulatory response. By contrast, the previous process depended on a more subjective, qualitative, and deterministic assessment, primarily of inspection findings to assess licensee performance according to a numerical rating (1 to 3 in decreasing order of performance) in four areas of performance (operations, maintenance, engineering and plant support). The new assessment process also features expanded reviews on a semiannual basis, including inspection planning and a performance report.

6.2.2.7 Experience and Examples

The NRC established a pilot program in 1999 for nine plants applying the new oversight process to test its effectiveness and to identify possible problems. The agency began to fully implement the process in April 2000.

On average, the NRC staff expended about 3250 inspection hours at each operating reactor during the year 2000. The NRC has adjusted the total direct onsite inspection hours on the basis of each nuclear power plant's operating experience in recent years. Performance at reactors has been improving, and this is reflected in the downward trend of the total hours across the inspection program. To ensure that the NRC effectively allocates resources to maintain reactor safety, the agency retains significant flexibility to conduct additional inspections of potentially significant events and plant conditions.

One plant-specific example is the following. A steam generator tube leak at Indian Point, Unit 2, on February 15, 2000, resulted in declaration of an "ALERT." This event occurred while the NRC was in transition to the new oversight process. A reactive NRC inspection team responded to the site to evaluate licensee actions and to obtain information needed to determine the significance of the event. Subsequent inspection findings (and performance indicator data) from this event were assessed under the new process. For instance, steam generator tube leakage is reported by the reactor coolant system leakage performance indicator under the Barrier Integrity cornerstone. The licensee reported a "yellow" coding from this indicator. The NRC inspections subsequently identified findings in emergency preparedness as "white." The NRC also determined that the licensee's overall direction and execution of inservice inspections of the steam generator in 1997 was "red" because, despite opportunities, the licensee did not identify and correct a significant condition adverse to quality (namely the presence of flaws caused by primary water stress corrosion cracking). The licensee did not adequately account for conditions that adversely affected the detectability of flaws, and increased the susceptibility of the tubes. As a result, flawed tubes were left in service after this inspection until the failure on February 15, 2000. As discussed above, such findings are then considered in inspection planning. The NRC recently issued an inspection report on its followup inspection efforts. That report is available at the agency's web site at *http://www.nrc.gov*. This example showed that the new program retained a very responsive reaction to a potentially significant performance problem.

6.2.3 Accident Sequence Precursor Program

6.2.3.1 Description of the Program

The Accident Sequence Precursor Program views U.S. nuclear plant operating experience from a perspective of safety significance. The primary aim of the program is to systematically identify, document, and rank operating events that are the most significant in potential to cause inadequate core cooling and core damage. The program has the secondary objectives of (1) categorizing precursors for plant-specific and generic implications, (2) providing a measure to trend the risk of core damage, and (3) providing a partial check on dominant core damage scenarios predicted by probabilistic risk analyses.

An accident sequence precursor is an historically observed element or condition in a postulated sequence of events leading to some undesirable consequence. For purposes of the program, the undesirable consequence is usually severe core damage. Applying probabilistic risk techniques, the staff evaluates initiating events, degradation of plant conditions, and failures of safety equipment that could increase the probability of postulated accident sequences. Accident sequences of interest are those that, if additional failures occurred, would have resulted in inadequate core cooling that could have caused severe core damage. For example, a loss-of-coolant accident with a reported failure of a high-pressure injection system may be postulated. In this example, the precursor would be the failure of the high-pressure injection system.

To identify potential precursors, the staff reviews licensee event reports or other documents (such as inspection reports and reports by incident investigation teams of plant problems, equipment failures, or other operational incidents). The staff analyzes events it considers to be potential precursors and calculates the conditional probability (i.e., probability, given an initiating condition) of core damage by mapping failures observed during the event onto the program's core damage models. The staff identifies and documents events with conditional probabilities of subsequent severe core damage greater than 1×10^{-6} as precursors. It considers precursors with values greater than 1×10^{-4} important.

The Accident Sequence Precursor Program began in 1979. Since then, the staff has evaluated and documented about 600 precursors from reported experience for 1969 through 1999. The precursor events identified by the Accident Sequence Precursor Program form a unique database of historical system failures, multiple losses of redundancy, and infrequent core damage initiators. The program has evolved to the point where the NRC routinely uses the methodology and results. The NRC continues to improve the methodology to better account for plant design and operational differences, human reliability, and changes in equipment, and to provide user-friendly analytical tools. Planned improvements include incorporating modeling and data uncertainty into each event analysis, establishing a more complete set of accident sequences, and better evaluating the containment response and consequences.

Commercial nuclear power reactors in the United States now have a combined total of more than 2000 years of operating experience. The Accident Sequence Precursor Program uses information gained from this experience to continually assess plant operation. The assessment helps to determine how well plant designs and capabilities can cope with actual operational events or conditions.

6.2.2.2 Experience and Examples

NUREG/CR-4674, Volume 27, dated July 2000, describes the evaluation of events that occurred in 1998. The primary result of the Accident Sequence Precursor Program is the identification of operational events with a conditional core damage probability or an importance greater than or equal to 1×10^{-6} that satisfy at least one of the four precursor selection criteria: (1) a core damage initiator requiring safety systems response, (2) the failure of a system required to mitigate the consequences of a core damage initiator, (3) degradation of more than one system required for mitigation, or (4) a trip or loss of feedwater with a degraded mitigating system. Nine of the at-power events that occurred during 1998 were found to have a conditional core damage probability or an importance greater or equal to 1×10^{-6} and were sent to the respective licensees for comment. The 1998 events revealed no shutdown-related precursors. The tables shown in Annex 2 list the accident sequence precursors for 1998.

The category, "Potentially Significant Events Considered Impractical to Analyze", contained no potentially significant events for 1998. Typically, this category includes events that are impractical to analyze because of a lack of information or inability to reasonably model the event within a probabilistic risk analysis framework, given the level of detail typically available in the models used.

One containment-related event was identified for 1998. That event is listed in Table 3.5, " Index of "Containment-Related" Events" shown in Annex 2. This category includes losses of containment functions, such as containment cooling, containment spray, or hydrogen control.

Two "interesting" events were identified for 1998. They are listed in Table 3.6, "Index of "Interesting" Events" shown in Annex 2. This category includes events that were not designated as precursors, but give insight into unusual failure modes that could compromise continued core cooling.

Precursors with conditional core damage probabilities greater than or equal to 1×10^{-4} have traditionally been considered important by the Accident Sequence Precursor Program. In 1998, a tornado touchdown at Davis-Besse was identified as a precursor with a conditional core damage probability greater than or equal to 1×10^{-4}. The Davis-Besse Plant was in Mode 1 at 99% power on June 24, 1998, when a severe thunderstorm cell moved into the area. Several minutes later, a tornado touched down either near or in the switchyard, damaging switchyard equipment and causing a complete loss-of-offsite power. Before the touchdown of the tornado, the senior reactor operator instructed the operators to start the emergency diesel generators from the control room because of the severe weather conditions. Although emergency diesel generator 2 started successfully, emergency diesel generator 1 failed to start. Operators then tried to start emergency diesel generator 1 locally and succeeded. Minutes later, a tornado touched down in or near the switchyard, causing a complete loss of offsite power. This loss-of-offsite power caused the turbine control valves to close in response to a load rejection by the main generator. The reactor protection system initiated a reactor trip on high reactor coolant system pressure. The licensee declared an ALERT as prescribed by the plant's emergency procedures. About 1 day later, after an offsite power line was restored, the emergency diesel generators were shut down, and the ALERT was subsequently downgraded. The conditional core damage probability for this event is about 5.6×10^{-4}.

The total number of precursors identified each year since 1988 is shown in Figure 3.1, Annex 3. Results of recent years cannot be directly compared with those of earlier years without substantial effort to reconcile differences in analysis. Major differences in selecting and modeling events were implemented for event assessments for 1984, 1988, 1992, 1994, and 1998. In particular, the conditional core damage probabilities estimated for some 1992--1998 precursors are lower than those for equivalent precursors in earlier years because the analyses incorporated supplemental and plant-specific mitigating systems beyond those that were included in the pre-1992 models. In addition, new modeling techniques, such as plant- specific fault trees, were adopted for the analysis of the 1994-1998 precursors. Because of the differences in analysis methods, only limited observations are given here, as follows. The number of precursors for 1998 (9) is about double the number from 1997 (5) but appears to be about the average of the post-1992 years. Some of the statistically significant downward trend shown in Figure 3.1 of Annex 3 over a 10-year period may be ascribed to differences in the approach to analysis.

6.2.4 Program for Resolving Generic Issues

The program for resolving generic issues is described in NUREG-0933, "A Prioritization of Generic Safety Issues." The most recent version of NUREG-0933, published in June 2000 and incorporating Supplements 1 through 24, is available on the NRC Website at *http://www.nrc.gov.*

The sources of potential generic issues include safety-related research, risk-assessment analyses, and public and industry concerns. Major safety issues affecting multiple facilities, and addressed in the Program for Resolving Generic Issues are the Three Mile Island (TMI) Action Plan requirements, unresolved safety issues, generic safety issues, and all other multi-plant actions.

To prioritize each issue, the NRC has developed a benefits assessment method that is predicated on risk. In prioritizing issues, the NRC considers both operating and future plants. The NRC identifies issues that could improve safety and the resolution cost. The NRC prioritizes such issues to allocate resources to all of the safety issues that have a high potential for reducing risk, and to not further consider those that have little safety significance and little promise of significantly enhancing safety. However, the NRC does not prioritize issues that demand immediate action because it must quickly make decisions on such issues. Generally, "immediate action" means issuing a Bulletin or an Order (types of NRC instruments that result in requirements).

Prioritization of activities primarily focuses on generic safety issues (i.e., safety issues that may affect all, several, or a class of nuclear power plants, and that may lead to improvements in safety and issuing of new or revised requirements or guidance). However, the NRC's way of prioritizing has been and can be used to identify changes in requirements that could significantly reduce the effect (usually the cost) on licensees without substantially changing public risk. The NRC classifies issues of this type as regulatory impact issues to clearly differentiate them from those that improve the safety of nuclear power plants but may be still worthwhile.

The following groups of issues are subject to prioritizing:

- TMI Action Plan items in NUREG-0660, "NRC Action Plan Developed as a Result of the TMI-2 Accident."

- Task Action Plan items identified in NUREG-0371, "Approved Category A Task Action Plans," and NUREG-0471, "Generic Task Problem Description, Category B, C and D Tasks," as well as the issues A-42 through A-49 that were designated as unresolved safety issues in NUREG-0471.

- New generic issues identified by the staff, the Advisory Committee on Reactor Safeguards, and others.

- Human Factors Program Plan items in NUREG-0985,"Nuclear Regulatory Commission Human Factors Program Plan, Revision 2."

- Chernobyl issues in NUREG-1251, "NRC Action Plan Developed as a Result of the TMI-2 Accident, Volume 2."

The NRC considers an issue resolved when its review has resulted in either (1) establishing regulatory requirements or guidance (by rule, standard review plan change, or equivalent) or (2) a documented authoritative decision that no change in requirements is warranted.

All TMI Action Plan items, Task Action Plan items, new generic issues, and human factors issues included in the program and their priority are listed in Table II of NUREG-0933.

6.2.5 Rulemaking

The NRC publishes formal requirements regarding reactor safety in Title 10 of the Code of Federal Regulations (10 CFR). The agency updates these regulations frequently for a variety of reasons. When the agency identifies a safety issue that requires a change in the regulations, it undertakes a rulemaking action in accordance with the process established by the Administrative Procedures Act. Typically, the NRC publishes a proposed rule for public comment, addresses the comments, and then issues the final rule. The process usually takes at least 18 months from the start of the rulemaking action until the publishing of the final rule. The agency uses other mechanisms, such as generic communications, discussed in Article 19, Section 19.7, when faster action is needed.

Examples of major rules the agency has developed to upgrade its formal requirements regarding reactor safety are 10 CFR 50.65, "Requirements for Monitoring the Effectiveness of Maintenance at Nuclear Power Plants," (1991) and 10 CFR 50.63, "Loss of All Alternating Current Power" (1988).

6.2.6 Decommissioning

When a licensee decides to shut down a plant, it must notify the NRC in writing within 30 days and submit a Post-Shutdown Decommissioning Activity Report. The report must address planned decommissioning activities, and contain a schedule of significant milestones and documentation that demonstrates that environmental impacts have been considered. Subsequently, a licensee must submit a license termination plan within 2 years before the expected license termination, addressing in detail the final radiation release, site characterization and remediation, and any new information. Before the NRC approves the plan, it provides an opportunity for a hearing and holds a public meeting near the plant.

Licensees may choose one of three methods for decommissioning their plants: DECON, SAFSTOR or ENTOMB. Under DECON (immediate dismantlement), soon after closing the nuclear plant, the licensee removes equipment, structures, and parts of the plant containing radioactive contaminants or decontaminates them to a level that permits release of the property and termination of the NRC license. Under SAFSTOR, the licensee maintains a nuclear plant in a condition that allows the radioactivity to decay, monitors the plant, and later dismantles it. Under ENTOMB, a licensee encases radioactive contaminants in a structurally sound material, such as concrete, and properly maintains and monitors the structure until the radioactivity decays to a level permitting release of the property.

In decommissioning special care is needed in the handling of reactor parts and plant components that have become radioactive during decommissioning work, and contaminated materials must be shipped to a low-level radioactive waste disposal site for burial. The NRC has adopted extensive regulations to ensure that decommissioning is accomplished safely and that residual radioactivity is reduced to a level that permits release of the property for unrestricted use or under restricted conditions. The NRC oversees the decommissioning of nuclear reactors through inspections that emphasize radiological controls, management, compliance with procedures, spent fuel pool operations, and the safety review program. During decommissioning, the NRC reviews and approves license termination plans, conducts inspections, processes license amendments, and monitors the status of activities to ensure that radioactive contamination is reduced or stabilized. This monitoring ensures that safety requirements are being met throughout the process. To be acceptable, decommissioning must be completed generally within 60 years. The NRC will consider a period longer than 60 years only when necessary to protect public health and safety.

6.2.7 Research Program

The NRC conducts reactor safety research to support its mission of ensuring that its licensees safely design, construct, and operate nuclear reactor facilities. The agency carries out this research program to identify, evaluate, and resolve safety issues, to ensure that an independent technical basis exists to review licensee submittals, to evaluate operating experience and results of risk assessments for safety implications, and to support the development and use of risk-informed regulatory approaches. In conducting the Reactor Safety Research program, the NRC will anticipate challenges posed by the introduction of new technologies and changing regulatory demand. The NRC continues to seek out opportunities to leverage its resources through domestic and international cooperative programs, and provide enhanced opportunities for stakeholder involvement and feedback on its research program. In addition, the agency will conduct research to address technical issues that it anticipates will arise during its review of advanced reactor designs. The Reactor Safety Research program is directly aligned with the NRC's performance goals.

6.2.8   Programs for Public Participation, Handling Petitions and Allegations, and Resolving Differing Professional Views and Opinions

The NRC views building and maintaining public trust and confidence that it is carrying out its mission as an important element of its safety responsibility. Toward this end, the NRC provides the diverse body of stakeholders (general public, Congress, other Federal, State and local governments, Indian Tribes, industry, technical societies, the international community and citizen groups) clear and accurate information about its role, and opportunities to participate in the agency's regulatory programs. The NRC's goal of increasing public confidence has resulted in various programs that give the public more accessibility, make communication with stakeholders more clear, accurate, reliable, objective and timely, and reporting of the performance of nuclear power facilities open and objective. The NRC has developed web sites to disseminate timely, accurate information regarding issues of interest to the public or events at nuclear facilities. The NRC elicits public involvement early in the regulatory process so as to address any safety concerns in a timely manner. In addition to the formal petition and hearing processes integrated into the licensing program, the agency uses feedback forms at public meetings to obtain public input.

Fostering an environment in which safety issues can be openly identified without fear of retribution is of paramount interest to the NRC. Examples of ways used by the NRC to identify safety concerns include the petition process under 10 CFR 2.206, "Requests for Action under this Subpart," the safety conscious work environment program, the allegation program, and the program for Differing Professional Views and Opinions.

The NRC regulations in 10 CFR 2.206 allow any member of the public to raise potential health and safety concerns and ask the NRC to take specific enforcement actions against a licensee. If warranted, the NRC can modify, suspend or revoke a license or take other appropriate enforcement action to resolve a problem identified in the petition. Recent changes made to the petition process emphasize a timely response to the petitioner and encourage increased, direct involvement of the petitioner (in addition to involvement of the licensee) by allowing the petitioner to personally address the petition review board and comment on the agency decision.

The NRC encourages workers in the nuclear industry to take their concerns directly to their employers (often the licensee or plant operator) and is particularly vigilant about a safety conscious work environment that encourages such reporting. The NRC programs also ensure that there is no fear of retribution of licensee action following such reporting. Additionally, the workers or any member of the public can bring their safety concerns directly to the NRC. The agency established a toll-free safety hotline for reporting such safety concerns, and NRC management, staff and inspectors, including the resident inspectors at plant sites, are trained and available to receive such concerns. (See following discussion on allegation program.)

Historically, 1000 to 2300 potential safety or regulatory issues have been reported directly to the NRC each year by industry workers or members of the public. The NRC developed the allegation program to establish a formal process for evaluating and responding to each issue. The primary purpose of the program is to provide an alternative method for individuals to raise safety or regulatory issues and to have them resolved. About seventy percent of the issues that are reported to the NRC are from licensee employees, employees of contractors to licensees, or former employees of licensees or contractors. With sufficient information, the NRC will evaluate each issue to determine whether it can verify the issue and, if so, the effect of the issue on plant safety. The evaluation is either an engineering review, inspection or investigation by the NRC staff or an evaluation by the licensee that is reviewed by the NRC staff. Historically, the NRC has been able to verify twenty five to thirty percent of the issues received. If the evaluation reveals a violation of regulatory requirements, the agency takes appropriate enforcement actions. Additionally, the NRC informs the individual who raised the issue of the results of its evaluation in writing.

The NRC's programs for Differing Professional Views and Differing Professional Opinions allow NRC employees to bring safety concerns or other issues important to the mission of the NRC to senior NRC management and, where appropriate, to the Commission. A panel of experts reviews the concern to determine the need for regulatory action.

ARTICLE 7. LEGISLATIVE AND REGULATORY FRAMEWORK

1. **Each Contracting Party shall establish and maintain a legislative and regulatory framework to govern the safety of nuclear installations.**

2. **The legislative and regulatory framework shall provide for:**

    (i)   the establishment of applicable national safety requirements and regulations

    (ii) a system of licensing with regard to nuclear installations and the prohibition of the operation of a nuclear installation without a license

    (iii) a system of regulatory inspection and assessment of nuclear installations to ascertain compliance with applicable regulations and the terms of licences

    (iv) the enforcement of applicable regulations and of the terms of licences, including suspension, modification, and revocation

This section explains the legislative and regulatory framework governing the U.S. nuclear industry. It discusses the provisions of that framework for establishing national safety requirements and regulations, and systems for licensing, inspection, and enforcement.

7.1 Legislative and Regulatory Framework

The Atomic Energy Act of 1954, passed by Congress and signed by the President, provided the framework for all subsequent regulation of nuclear installations. However, as is generally the case with most laws, this Act provided general principles and concepts, and left it up to the regulatory body (the NRC) to address the details by issuing specific regulations.

7.2 Provisions of the Legislative and Regulatory Framework

7.2.1 National Safety Requirements and Regulations

The regulations on nuclear installations in the U.S. are governed by the Atomic Energy Act of 1954 (as amended) and the Energy Reorganization Act of 1974. (The Energy Reorganization Act abolished the U.S. Atomic Energy Commission, and created the NRC and the U.S. Energy Research and Development Administration (ERDA). ERDA was subsequently incorporated into the U.S. Department of Energy (DOE)). The NRC administers these statutes in licensing nuclear installations in the U.S. In addition, the following statutes bear substantially on the practices and procedures of the Commission:

*   Administrative Procedure Act
*   Chief Financial Officers Act of 1990
*   Clean Air Act of 1977
*   Coastal Zone Management Act

- Comprehensive Environmental Response, Compensation, and Liability Act of 1980
- Endangered Species Act
- Federal Advisory Committee Act
- Federal Water Pollution Control Act–also known as the Clean Water Act of 1977
- Freedom of Information Act
- Government in the Sunshine Act
- Government Paperwork Elimination Act
- Inspector General Act
- Low-Level Radioactive Waste Policy Amendments Act of 1985
- National Environmental Policy Act of 1969
- National Historic Preservation Act
- Nuclear Waste Policy Act of 1982
- Price Anderson Act of 1957
- Privacy Act
- Resource Conservation and Recovery Act of 1976
- Toxic Substances Control Act of 1976
- Uranium Mill Tailings Radiation Control Act of 1978
- West Valley Demonstration Project Act of 1980
- Wild and Scenic Rivers Act

Nuclear installations in the U.S. must be licensed by the NRC. (Some government facilities that are operated by or for the DOE, are exempt from licensing by the requirements of the Atomic Energy Act and the Energy Reorganization Act.) The following facilities must be licensed:

- Nuclear reactors (power, test, and research)
- Uranium mills
- Solution recovery plants (milling)
- Uranium dioxide (UO_2) and mixed oxide (MOX) fuel fabrication plants
- Spent fuel storage (interim)
- High-level waste and spent fuel geologic repositories
- Low-level waste burial grounds
- Fuel reprocessing plants
- Isotopic separation (enrichment) plants

Rules and regulations governing the licensing of these facilities are set down in Title 10 of the *U.S. Code of Federal Regulations* (10 CFR). The first 200 parts of 10 CFR (Parts 0 -199) apply to the NRC.

Although the licensing process is similar for reactors, separation facilities, reprocessing plants, and nuclear waste storage and disposal facilities, the following discussion describes the licensing practices for nuclear power plants.

7.2.2 Licensing of Nuclear Installations

Chapter 10, Section 103, "Commercial Licenses" of the Atomic Energy Act of 1954" grants the NRC authority to issue licenses for nuclear reactor facilities. In addition, Section 103 states that such licenses are subject to such conditions as the NRC may by rule or regulation establish to effectuate the purposes and provisions of the Atomic Energy Act. Section 103.b states the following:

> The Commission shall issue such licenses on a nonexclusive basis to persons applying therefore (1) whose proposed activities will serve a useful purpose proportionate to the quantities of special nuclear material or source material to be utilized; (2) who are equipped to observe and who agree to observe such safety standards to protect health and to minimize danger to life or property as the Commission may by rule establish; and (3) who agree to make available to the Commission such technical information and data concerning activities under such license as the Commission may determine necessary to promote the common defense and security and to protect the health and safety of the public.

All such information may be used by the Commission only for the purposes of the common defense and security and to protect the health and safety of the public.

Section 182, "License Applications," states the following:

> Each application for a license hereunder shall be in writing and shall specifically state such information as the Commission, by rule or regulation, may determine to be necessary to decide such of the technical and financial qualifications of the applicant, the character of the applicant, the citizenship of the applicant or any other qualifications of the applicant as the Commission may deem appropriate for the license.

Section 189.a of the Atomic Energy Act provides affected parties with hearing rights in proceedings for the granting, suspending, revoking, or amending of a license or construction permit. Formal hearings, which are used in licensing proceedings for production and utilization facilities (e.g., nuclear power plants), are held under procedural rules stated in 10 CFR Part 2, "Rules of Practice for Domestic Licensing Proceedings and Issuance of Orders," Subpart G, "Rules of General Applicability." The staff participates as a party in all formal hearings, and may also participate as a party in informal hearings. Subpart G hearings are usually held before a three-member Atomic Safety and Licensing Board, which is generally composed of one lawyer and two technical members.

A party may appeal the decision of a Licensing Board to the Commission itself. Review by the Commission is discretionary, not mandatory. The Commission ordinarily reviews only novel or complex issues or important legal, policy, or health and safety questions. In deciding whether to grant a review of a particular case, and in writing its appellate decisions, the Commission is assisted by the Office of Commission Appellate Adjudication.

The licensing process is described in greater detail in Article 18. There are two alternative approaches. The traditional approach under 10 CFR Part 50, "Domestic Licensing of Production and Utilization Facilities," requires two steps. The NRC reviews a preliminary application and grants a construction permit, and later reviews the final application and grants an operating license. All current operating plants in the U.S. were licensed according to this process.

In 1989, the Commission established an alternative licensing system, published in 10 CFR Part 52, "Early Site Permits; Standard Design Certifications; and Combined Licenses for Nuclear Power Plants," that provides for certified standard designs and combined licenses that resolve design issues before construction, and early site permits that resolve siting issues before construction. The basic concept underlying 10 CFR Part 52 is that nuclear reactor designs can be approved through generic rulemaking. Once approved, an applicant can use them in applications for permission to build and operate nuclear power plants without the necessity to relitigate, in individual hearings, the issues resolved in the rulemaking. Moreover, the criteria for determining whether the plant had been built as specified would be determined and approved before construction. Thus completed, the plant could begin operation without a second hearing, provided that it satisfied the acceptance criteria. To the extent possible, issues would be litigated before construction, not after construction is largely complete, when the consequences of delay are much greater. In adopting 10 CFR Part 52, the Commission used the latitude allowed by law to streamline licensing.

The Energy Policy Act of 1992 codified major parts of 10 CFR Part 52 and directed the NRC to issue implementing regulations. The legislation provided that the Commission may allow a completed plant to operate during the pendency of any post-construction hearing, provided that certain safety findings can be made. In addition, the legislation authorized the Commission to use either formal or informal procedures for such post-construction hearings.

7.2.3 Inspection and Assessment

Under the Atomic Energy Act of 1954, the NRC has the authority to inspect nuclear power plants in its role of protecting public heath and safety. The NRC staff inspects power reactors under construction, in test conditions, and in operation to ascertain compliance with regulations and license conditions. Through its inspection program, the NRC assesses whether activities are properly conducted and equipment is properly maintained to ensure safe operations. The NRC integrates inspection results into its overall evaluation of licensee performance, as discussed in Article 6. When the NRC discovers a safety problem or failure to comply with requirements, it requires prompt corrective action by the licensee, reinforcing it, if necessary, by enforcement action.

7.2.4 Enforcement

The NRC's enforcement jurisdiction is drawn from the Atomic Energy Act of 1954, as amended, and the Energy Reorganization Act (ERA) of 1974, as amended.

Section 161 of the Atomic Energy Act authorizes the NRC to conduct inspections and investigations, and to issue Orders as may be necessary or desirable to promote the common defense and security or to protect health or to minimize danger to life or property. Section 186 authorizes the NRC to revoke licenses under certain circumstances (e.g., for material false statements, for a change in conditions that would have warranted NRC refusal to grant a license on an original application, for a licensee's failure to build or operate a facility in accordance with the terms of the permit or license, and for violation of an NRC regulation). Section 234 authorizes the NRC to impose monetary civil penalties not to exceed $100,000 per violation per day; but that amount is adjusted every four years by the Federal Civil Penalties Inflation Adjustment Act and is currently $120,000. In addition to the provisions mentioned in Section

234, Sections 84 and 147 authorize the imposition of civil penalties for violations of regulations implementing those provisions. Section 232 authorizes the NRC to seek injunctive or other equitable relief for violation of regulatory requirements.

Section 206 of the Energy Reorganization Act authorizes the NRC to impose civil penalties for knowing and conscious failures to provide certain safety information to the NRC.

Chapter 18 of the Atomic Energy Act provides for varying levels of criminal penalties (i.e., monetary fines and imprisonment) for willful violations of the Act and regulations or Orders issued under Sections 65, 161(b), 161(i), or 161(o) of the Act. Section 223 provides that criminal penalties may be imposed on certain individuals who are employed by firms constructing or supplying basic components of any utilization facility if the individual knowingly and willfully violates NRC requirements in a manner that could significantly impair a basis component. Section 235 provides that criminal penalties may be imposed on persons who interfere with nuclear inspectors. Section 236 provides that criminal penalties may be imposed on persons who attempt to or cause sabotage at a nuclear facility or to nuclear fuel. Alleged or suspected criminal violations of the Atomic Energy Act are referred to the U.S. Department of Justice for appropriate action.

The procedures that the NRC uses in exercising its enforcement authority are specified in 10 CFR Part 2,"Rules of Practice for Domestic Licensing Proceedings and Issuance of Orders," Subpart B, "Procedure for Imposing Requirements by Order, or for Modification, Suspension, or Revocation of a License, or for Imposing Civil Penalties." The scope of Subpart B includes the procedures described below.

10 CFR 2.201, "Notice of Violation," states the procedure for issuing Notices of Violation .

10 CFR 2.202, "Orders," states the procedure for issuing Orders. In accordance with 10 CFR 2.202, the NRC may decide to issue an Order to institute a proceeding to modify, suspend, or revoke a license or to take other action against a licensee or other person subject to the jurisdiction of the NRC. The licensee or any other person adversely affected by the order may request a hearing. The NRC is authorized to make Orders immediately effective if required to protect public health, safety, or interest, or if the violation is willful.

10 CFR 2.204, "Demand for Information," specifies the procedure for issuing a Demand for Information to a licensee or other person subject to the Commission's jurisdiction to determine whether an Order should be issued or other enforcement action should be taken. The Demand does not provide hearing rights, as only information is being sought. A licensee must answer a Demand. An unlicensed person may answer a Demand by either providing the requested information or explaining why the Demand should not have been issued.

10 CFR 2.205, "Civil Penalties," states the procedure for assessing civil penalties. The NRC initiates the civil penalty process by issuing a Notice of Violation and proposed imposition of a civil penalty. The agency provides the person charged with an opportunity to contest in writing the proposed imposition of a civil penalty. After evaluating the response, the NRC may mitigate, remit or impose the civil penalty. If the agency imposes a civil penalty, it provides an opportunity for a hearing. If a civil penalty is not paid following a hearing, or if a hearing is not requested, the

agency may refer the matter to the U.S. Department of Justice to institute a civil action in Federal District Court to collect the penalty.

ARTICLE 8. REGULATORY BODY

1. **Each Contracting Party shall establish or designate a regulatory body entrusted with the implementation of the legislative and regulatory framework referred to in Article 7, and provided with adequate authority, competence, and financial and human resources to fulfill its assigned responsibilities.**

2. **Each Contracting Party shall take the appropriate steps to ensure an effective separation between the functions of the regulatory body and those of any other body or organization concerned with the promotion or utilization of nuclear energy.**

This section explains the establishment of the U.S. regulatory body, the NRC. It also explains how the functions of the NRC are separate from those of bodies responsible for promoting and using nuclear energy (i.e., the U.S. Department of Energy).

8.1 The Regulatory Body

This section explains the NRC's mandate, authority and responsibilities, status as an independent regulatory agency, structure, and staffing and resources

8.1.1 Mandate

As discussed in Article 7, the U.S. NRC was created as an independent regulatory agency in January 1975, with the passage of the Energy Reorganization Act of 1974. In giving the NRC an exclusively regulatory mandate, the statute reflected (in part) a Congressional judgment that the commercial nuclear power industry (which, at that time, was expanding rapidly and was expected to grow still more) had reached a point at which the full-time attention of an exclusively regulatory agency was warranted. A second major reason for the NRC's creation was a developing public concern that regulatory responsibilities were overshadowed by the promotion of nuclear power at the Atomic Energy Commission. By 1975, a large segment of the U.S. population was skeptical about the role of nuclear power in the Nation's energy mix for a variety of reasons. Specifically, these reasons included the safety of nuclear plants, security concerns (including the threat that terrorists would attack nuclear facilities or make weapons from stolen nuclear materials), the proliferation of nuclear weapons-making capacity around the world, and the lack of a solution to the problem of high-level radioactive waste disposal. It was hoped that by limiting the NRC responsibilities to regulation, making it an independent agency, and choosing the multi-member bipartisan Commission format, greater public confidence would ensue in governmental decision making in the nuclear area.

8.1.2 Authority and Responsibilities

This section discusses the scope of authority of the NRC, its international responsibilities, and status as an independent regulatory body.

8.1.2.1 Scope of Authority

The NRC's mission is to ensure that the civilian uses of nuclear energy and materials in the U.S. are conducted with proper regard for public health and safety, national security, environmental concerns, and (in the case of the initial licensing of nuclear power plants) the antitrust laws. The basic charter for these regulatory responsibilities is the Atomic Energy Act of 1954, through which Congress created a national policy of developing the peaceful uses of atomic energy. That statute has been amended over the years to deal with developing technology and changing perceptions of regulatory needs. For example, antitrust reviews were added in 1970, the same year that the National Environmental Policy Act imposed broad new responsibilities on Federal agencies. Other more specialized statutes prescribe the NRC's duties with regard to high-level radioactive waste, low-level radioactive waste, mill tailings, environmental reviews, nonproliferation, anti-terrorism, and import/export.

The Atomic Energy Act of 1954 has been described by the courts as "almost unique" in the degree of discretion that it confers on the NRC to make the decisions it believes to be best, using the procedures it considers most suitable. Only very rarely will an NRC decision be overturned on substantive grounds. The language of the Act granting the NRC discretionary authority contrasts with the prescriptive language of later statutes governing such agencies as the Environmental Protection Agency and the Occupational Safety and Health Administration.

The NRC's licensing authority extends to other Federal agencies (such as the Tennessee Valley Authority, which operates nuclear power plants) and the military, which uses radiopharmaceuticals in its hospitals. The NRC's responsibilities include both safety and "safeguards" through which the agency ensures the security of nuclear facilities and materials against radiological sabotage and thefts.

In addition, the NRC is authorized to delegate certain regulatory authorities to Agreement States for most nuclear materials (although not nuclear power plants) and fuel facilities of Federal agencies. More than half of the States are Agreement States, and together they administer another 16,000 materials licenses.

The NRC has no authority to regulate either naturally occurring or accelerator-produced radioactive materials, other than uranium and thorium; these are regulated by either the Environmental Protection Agency or individual States. The NRC also has no authority to regulate machine-produced radiation, such as the emissions from X-ray units or linear accelerators; these units are regulated by either the Food and Drug Administration or individual States.

For the scope of authority over DOE nuclear installations, see Section 8.2 below.

8.1.2.2 International Responsibilities

Internationally, the NRC has a substantial program of cooperation with about 30 countries to exchange information relevant to the safety of nuclear facilities in the U.S. and abroad. The NRC also assists regulatory agencies in selected countries in strengthening their nuclear safety programs. For example, using funding by the Agency for International Development, U.S. Department of State, the Commission is providing assistance to the countries of the Former

Soviet Union (principally Russia and Ukraine) and the countries of Central and Eastern Europe to upgrade the safety of their civilian reactors programs.

Another key international responsibility is licensing the export and import of nuclear materials and equipment, such as low-enriched uranium fuel for nuclear power plants, high-enriched uranium for research and test reactors, nuclear reactors themselves, and certain nuclear reactor components (such as pumps and valves). The NRC's principal role is to ensure that such exports are consistent with the goal of limiting the proliferation of nuclear weapons. Standards and procedures for the issuance of export licenses are detailed in the Nuclear Non-Proliferation Act of 1978.

8.1.2.3 The NRC as an Independent Regulatory Agency

The Commission's status as an independent regulatory agency within the Executive Branch of the Federal Government means that its regulatory decisions cannot ordinarily be dictated by the President. (By law, however, the U.S. Office of Management and Budget reviews the proposed NRC budget.) Likewise, the Congress cannot override the Commission's decisions, except by duly enacted legislation. Executive Order 12866, "Regulatory Planning and Review," which provides for review of agency regulations by the U.S. Office of Management and Budget, exempts independent regulatory agencies from its coverage, and the NRC has historically resisted efforts by the Office of Management and Budget to secure for itself the power to review proposed NRC regulations for substance. The subject of the NRC's relationship to other governmental bodies is discussed in greater detail below.

The obligation of other entities not to interfere with the NRC's decision-making process implies a matching responsibility on the part of the Commissioners and their personal staffs to keep the decision process free from improper outside influence. This is especially important in the case of formal adjudications, which are governed by hearing procedures similar to those used in trials before a judge in Federal court. When the Commissioners take part in formal adjudications, they ordinarily act in the role of appellate judges (reviewing the decisions of lower judges) and, in general, are bound by the same kinds of strictures applicable to Federal judges.

8.1.3 Structure of the Regulatory Body

This section explains the structure of the NRC. It covers the Commission, component offices and their responsibilities, and advisory committees and their functions.

8.1.3.1 The Commission

The Energy Reorganization Act of 1974 requires that the five Commissioners be U.S. citizens, and that no more than three belong to the same political party. Commissioners serve for fixed five-year terms, and are removable only for cause. The Chairman is designated by the President, and serves at the President's pleasure. If the Chairman is relieved of that position by the President, he or she remains a Commissioner for the duration of the term. The statute further provides that each member of the Commission shall have full access to information that is necessary to fulfill the member's duties and "equal responsibility and authority in all decisions and actions of the Commission."

The Chairman has responsibilities, as spokesman and Chief Executive Officer of the Commission, that significantly differentiate the Chairman's role from that of the other Commissioners. The Chairman initiates the appointment, subject to the approval of the Commission, of the Executive Director for Operations, the General Counsel, the Secretary, the members of the Commission's adjudicatory panel, the Chief Financial Officer, the Chief Information Officer, and (in consultation with the Executive Director for Operations) the Directors of the Commission's major program offices. The Chairman or a Commissioner can initiate removal of a holder of one of these offices, subject to the approval of the Commission.

To transact business, a quorum of Commissioners (i.e., a minimum of three) is required to be "present." Decision is by a majority of those participating; Commissioners may vote, abstain (in which case they are regarded as participating), or decline to participate. To approve final rules and issue adjudicatory decisions, the Commissioners must meet in person; other types of business are handled by written vote sheets.

The results of Commission deliberations are set forth in rules, Orders, and "staff requirements memoranda," which record how Commissioners voted, describe the outcome, and typically direct some unit of the staff to take a particular action. Most Commission actions are taken in response to NRC staff papers ("SECY papers"), which are generally made public after the Commission has made its decision.

In formal adjudicatory proceedings, the Commission is the ultimate decision-making body. For these purposes, the Commissioners occupy the role of judges, with all of the constraints that term implies on their freedom to discuss or gather information about a pending case. Like other judges, they must decide cases solely on the basis of the record before them, setting aside any knowledge of the issues gained through extra-judicial means, such as the media.

8.1.3.2 Component Offices of the Commission

Having needed to update regulatory programs, the NRC has extensively reorganized over the past several years. Toward that end, the agency realigned three major NRC program offices: the Office of Nuclear Reactor Regulation, the Office of Nuclear Materials Safety and Safeguards, and the Office of Nuclear Regulatory Research. It also eliminated the Office for Analysis and Evaluation of Operational Data, and reduced the number of NRC managers, as well as overall staffing and resource levels to improve overall organizational efficiency.

Executive Director for Operations

The position of the Executive Director for Operations is established by statute. As the head of the NRC staff, the director reports to the Chairman and is subject to the Chairman's supervision and direction. The director supervises and coordinates policy development and operational activities of the NRC's three major program offices. These duties also cover the NRC's regional offices and such other offices as those of Enforcement and Human Resources.

Office of Nuclear Reactor Regulation

The Office of Nuclear Reactor Regulation is one of the NRC's three major statutory program offices. This office is responsible for the licensing, inspection, and regulation of nuclear power and other reactors. Its reviews encompass the safety, safeguards, environmental, and antitrust aspects of these facilities, and it guides regional offices on facility inspections and licensing activities.

Office of Nuclear Material Safety and Safeguards

The responsibilities of the Office of Nuclear Material Safety and Safeguards include regulating the use and handling of nuclear and other radioactive materials; fuel fabrication and fuel development; medical, industrial, academic, and commercial uses of radioactive isotopes; safeguards activities; transportation of nuclear materials; out-of-reactor spent fuel storage; safe management and disposal of low-level and high-level radioactive wastes; uranium enrichment facilities; and management of related decommissioning. This office also guides regional offices on facility inspections and licensing activities. In addition, this office provides technical support to improve international safeguards on nuclear materials, which are administered by the International Atomic Energy Agency (IAEA).

Office of Nuclear Regulatory Research

The Office of Nuclear Regulatory Research plans, recommends, and carries out programs of nuclear regulatory research, standards development, and resolution of generic safety issues for nuclear power plants and other activities regulated by the NRC. In addition, this office is responsible for research confirming assessments of reactor safety, safeguards, waste management radiological protection, and environmental protection applying to the regulatory process.

Office of Enforcement

The Office of Enforcement develops policies and programs for enforcing the NRC 's requirements. In addition, this office approves Orders and civil penalties, and reviews other enforcement actions for consistency with the Commission's enforcement policy.

Office of the General Counsel

The Office of the General Counsel provides legal advice, opinions, and assistance to NRC's officials on all of the NRC's activities. It advises and assists the NRC's regional offices, although each region also has its own Regional Counsel, who is appointed by and reports to the Regional Administrator. This office interprets laws, regulations, and other sources of authority, and it advises on the legal form and content of proposed official actions. It represents the staff in licensing and enforcement proceedings; provides legal advice on enforcement actions; prepares, reviews, and interprets all of the NRC's contractual documents, interagency agreements, delegations of authority, regulations, orders, licenses, and other legal documents; reviews and gives opinions on issues of intellectual property, including patent, trademark, copyright and proprietary matters; and gives the NRC legal opinions and advice about the administration of the Freedom of Information, Privacy, and Government in the Sunshine Acts. It also advises the Commission on legislative matters and represents the NRC in the Federal Courts.

Office of the Chief Financial Officer

The Office of the Chief Financial Officer is responsible for the NRC's Planning and Budgeting, and Performance Management Process and for all of the NRC's financial management activities. This office establishes planning, budgeting, and financial management policy for the agency and provides advice to the Chairman and the Commission on these matters. This office develops and maintains an integrated agency accounting and financial management system; establishes policy and directs oversight of agency financial management personnel, activities, and operations; prepares and transmits an annual report which includes the agency's audited financial statement to the Chairman and the Director, Office of Management and Budget; monitors the financial execution of NRC's budget in relation to actual expenditures, controls the use of agency funds to ensure that they are expended in accordance with applicable laws and standards, and prepares and submits to the Chairman timely cost and performance reports; and reviews, on a periodic basis, fees and other charges imposed by NRC for services provided and makes recommendations for revising those charges as appropriate. The Office of the Chief Financial Officer provides an agency-wide management control program for financial and program managers to comply with the Federal Managers' Financial Integrity Act of 1982, and is responsible for implementing the Chief Financial Officers Act and the Government Performance and Results Act at the NRC.

Office of the Inspector General

The NRC's Inspector General reports to the Congress and the NRC Chairman; he or she is not subject to supervision by any other member of the NRC. The office's statutory responsibilities include (1) auditing and inspecting programs to promote effectiveness and efficiency in the operation of NRC's programs and (2) investigating allegations of wrongdoing by NRC employees and contractors to detect and prevent fraud, waste, and abuse.

Office of Investigations

The Office of Investigations investigates allegations of wrongdoing by NRC licensees and license applicants, as well as their contractors and vendors, in furtherance of the NRC's enforcement program. This office has the authority to initiate investigations, and to refer potential criminal violations directly to the Department of Justice without consultation with the Commission. This office does not investigate allegations against NRC employees and contractors; that is done by the Office of the Inspector General.

Office of State and Tribal Programs

The Office of State and Tribal Programs is responsible for establishing and maintaining good communication and working relationships between the NRC and other governmental entities, including the States, local governments, Native American tribes, and other Federal agencies. It administers the State Agreements Program by providing training and technical assistance and reviewing the adequacy and compatibility of the States' radiation control programs, which now have responsibility for some two-thirds of all materials licenses in this country.

Office of Congressional Affairs

The Office of Congressional Affairs is the primary contact point for all NRC communications with Congress. This office monitors legislative proposals, bills, and hearings, and informs the NRC of the views of Congress on NRC policies, plans, and activities.

Office of International Programs

The Office of International Programs formulates and carries out policies and programs on nuclear exports and imports, nonproliferation, international safeguards, international physical security, and international cooperation and assistance in nuclear safety and radiation protection. It maintains relationships with other Federal agencies, other countries, and such international agencies as the IAEA.

Office of Public Affairs

The Office of Public Affairs develops policies and programs for informing the public of the NRC's activities. Toward that end, this office disseminates information to the public and the news media concerning NRC policies, programs, and activities; keeps NRC management informed of media coverage; and directs and coordinates the activities of public information staffs located at regional offices.

The Office of Commission Appellate Adjudication

The Office of Commission Appellate Adjudication assists the Commission in its disposition of appeals of licensing board decisions and other adjudicatory matters coming before the Commission and monitors pending board cases. It also has lead responsibility for adjudication of certain aspects of license transfers.

Regional Offices

The NRC's four regional offices are located in the Philadelphia (Region I), Atlanta (Region II), Chicago (Region III), and Dallas (Region IV) areas. About 31 percent of the NRC's personnel are stationed in the regions. Each regional office is headed by a Regional Administrator. This administrator is responsible for executing established NRC policies and programs on inspection, enforcement, licensing, State agreements reviews, State liaison, and emergency response within the Region's boundaries.

The responsibilities of the NRC's regional offices include inspection and evaluation of engineering, construction, and operational activities of power reactors; operator licensing functions for power reactors; implementation of nuclear material safety licensing and inspection, emergency preparedness, and safeguards licensing functions assigned to the region; coordination of the NRC's Incident Response Program for activities within the region; issuance of Notices of Violation and proposed civil penalties (subject to further approval of Headquarters staff, depending on severity); review of Agreement State regulatory programs; and provision of technical assistance to Agreement States in carrying out their regulatory programs.

Office of the Secretary of the Commission

The Office of the Secretary of the Commission provides executive management services to support the Commission and to carry out Commission decisions. It advises and assists the Commission and staff on the planning, scheduling, and conduct of Commission business; maintains historical paper files of official Commission records, and administers the NRC Historical Program. The Secretariat maintains the Commission's official adjudicatory and rulemaking dockets including management of the Commission's Electronic Hearing Docket.

Support Offices

Supporting the Executive Director for Operations are the Offices of Administration, the Chief Information Officer, Human Resources, and Small Business and Civil Rights.

8.1.3.3 Advisory Committees and Licensing Boards

The Commission has four advisory committees chartered under the Federal Advisory Committee Act. This statute imposes certain constraints on advisory committees, primarily that they give advance notice of their meetings and, unless certain exemptions apply, hold them open to the public. The Commission also has licensing boards.

- Advisory Committee on Reactor Safeguards: Consisting of 10 members with expertise in scientific and engineering disciplines, this committee is the NRC's only statutory advisory committee. It advises on potential hazards of proposed or existing reactor facilities, the adequacy of proposed safety standards, and such other matters as the Commission may request. The statute requires that the Committee review certain types of applications, such as those for construction permits or operating licenses for power reactors or test reactors.

- Advisory Committee on Nuclear Waste: Consisting of 4 members with expertise in scientific and engineering disciplines, this committee advises on nuclear waste management issues as directed by the Commission.

- Advisory Committee on the Medical Uses of Isotopes: Consisting of 12 members, including qualified physicians and scientists and other representatives of the medical community, including a patients' representative, this committee considers medical questions referred by the NRC staff, and gives expert opinions on the medical uses of radioisotopes. It also advises the NRC staff, as required, on matters of policy. Members are employed under yearly personal services contracts.

- Licensing Support System Advisory Review Panel: This panel consists of 18 members who are representatives of the potential participants in NRC's future licensing proceeding for a high-level nuclear waste repository. The panel advises both the NRC and the U.S. Department of Energy on the Licensing Support System (now in the design stage), which is to be used in the proceeding to ensure that all pertinent documents are available to the participants.

- Atomic Safety and Licensing Boards: Adjudicatory hearings at the NRC are conducted by three-member licensing boards or a single presiding officer drawn from the Atomic Safety and Licensing Board Panel. These hearings mainly deal with nuclear reactor licensing, nuclear material licensing, and enforcement matters when licensees and other affected entities contest penalties or orders brought against them by the NRC staff for alleged infractions of the NRC's regulations. The Boards sometimes hold additional hearings on antitrust licensing, personnel matters, and special Commission-ordered proceedings.

8.1.4 Financial and Human Resources

This section discusses the budget and funding of the NRC, its human resources, and financial management.

8.1.4.1 Funding by Fees

Under the Omnibus Budget Reconciliation Act of 1990, 98 percent of the NRC's budget (less the costs for NRC work on the DOE High-Level-Waste Program, which are recovered from the DOE-administered Nuclear Waste Fund) must be recovered from NRC applicants and licensees through fees during fiscal year 2001 (October 1, 2000 to September 30, 2001). The Act was amended in 2000 to extend the requirement to collect fees through fiscal year 2005 and to reduce the amount of fees collected from 100 percent to 90 percent. Intended to mitigate the burden on NRC licensees to pay for costs for which they receive no direct benefit, the reduction is being phased in at two percent per year, beginning in fiscal year 2001, and continuing through fiscal year 2005.

Two types of fees are required to recover the NRC's budget authority. First, license and inspection fees, established by 10 CFR Part 170, "Fees for Facilities, Materials, Import and Export Licenses and Other Regulatory Services under the Atomic Energy Act of 1954, as Amended," recover the NRC's costs of providing individually identifiable services to specific applicants and licensees. The services provided by the NRC for which these fees are assessed are reviewing applications for the issuance of new licenses or approvals, amending or renewing licenses or approvals, and inspecting licenses. Second, annual fees, established by 10 CFR Part 171,"Annual Fees for Reactor Licenses and Fuel Cycle Licenses and Materials Licenses, Including Holders of Certificates for Compliance, Registrations, and Quality Assurance Program Approvals and Government Agencies Licensed by the NRC," recover generic (e.g., research and rulemaking) and other regulatory costs that are not recovered through 10 CFR Part 170 fees. The amounts of these fees are established each year through notice and comment rulemaking, and are predicated on the budget approved by Congress.

Tables 8.1 and 8. 2 below show the money and human resources to support the necessary NRC activities.

8.1.4.2 Financial Management

Since the Chief Financial Officers Act was passed in 1990, the NRC has carried out a number of actions to improve the agency's financial management. The Office of the Chief Financial Officer developed, coordinated, and published Management Directive and Handbook 4.2, "Administrative Control of Funds," and with the Office of Human Resources, has developed two training courses in financial management — Financial Management Seminar and Administrative Control of Funds Seminar — which are specifically tailored to the needs of NRC personnel.

Table 8. 1: Budget Authority by Appropriation

NRC Appropriation	FY 2000 Enacted	FY 2001 Enacted	FY 2002 Estimate	
			Request	Change from FY 2001
Salaries and Expenses (S&E) ($K)				
Budget Authority	464,913	481,825	506,900	25,075
Offsetting Fees	442,000	447,937	463,248	15,311
Net Appropriated—S&E	22,913	33,888	43,652	9,764
Office of the Inspector General (OIG) ($K)				
Budget Authority	5,000	5,500	6,180	680
Offsetting Fees	5,000	5,390	5,933	543
Net Appropriated—OIG	0	110	247	137
Total NRC ($K)				
Budget Authority	469,913	487,325	513,080	25,755
Offsetting Fees	447,000	453,327	469,181	15,854
Total Net Appropriated—NRC [2]	22,913	33,998	43,899	9,901

---

[2]

[1]Net Appropriation - NRC ($K)	FY 2000 Enacted	FY 2001 Enacted	FY 2002 Estimate
Nuclear Waste Fund	19,150	21,552	23,650
General Fund	3,763	12,446	20,249

Total	22,913	33,998	43,899

Table 8.2: Budget Authority and Staffing by Strategic Arena

	FY 2000	FY 2001	FY 2002 Estimate	
				Change from
Budget Authority by Strategic Arena				
Nuclear Reactor Safety	210,465	219,214	231,397	12,183
Nuclear Materials Safety	51,737	52,463	55,038	2,575
Nuclear Waste Safety	53,882	59,288	63,157	3,869
International Nuclear Safety Support	4,692	4,779	5,119	340
Management and Support	144,137	146,081	152,189	6,108
Subtotal (Salaries & Expenses)	**464,913**	**481,825**	**506,900**	**25,075**
Inspector General	5,000	5,500	6,180	680
Total NRC	**469,913**	**487,325**	**513,080**	**25,755**
Staffing (FTE) by Strategic Arena				
Nuclear Reactor Safety	1,430	1,424	1,425	1
Nuclear Materials Safety	399	377	382	5
Nuclear Waste Safety	259	266	271	5
International Nuclear Safety Support	39	38	39	1
Management and Support	630	614	617	3
Subtotal (Salaries & Expenses)	**2,757**	**2,719**	**2,734**	**15**
Inspector General	44	44	44	0
Total NRC	**2,801**	**2,763**	**2,778**	**15**
Reimbursable Business-Like FTE	13	11	11	0
Total NRC	**2,814**	**2,774**	**2,789**	**15**

8.1.5 Position of the NRC in the Governmental Structure

This section explains the relationship of the NRC to the Executive Branch, the States, and Congress.

8.1.5.1 Executive Branch

This section explains the relationship of the NRC to components of the Executive Branch. These components are the White House, Office of Management and Budget, U.S. Department of State, Environmental Protection Agency, Federal Emergency Management Agency, Department of Labor, and the U.S. Department of Justice.

The White House

As noted above, the NRC's status as an independent regulatory agency means that the White House cannot directly set NRC policy. It can, however, influence NRC policy by (1) appointing Commissioners and Chairmen in whose outlook and judgment it has confidence, and (2) making its views known on nonadjudicatory matters. In certain areas, such as national security policy, the Commission has declared its intent to give great weight to the views of the Executive Branch. In informal policy matters, such as rulemaking, White House and Executive Branch officials may properly try to influence NRC decisions, either publicly or privately; ultimately, however, the NRC must make the decision and take responsibility for it.

The NRC also works with the National Security Council and the White House in helping the U.S. develop policies for cooperating with and assisting the former Soviet Union.

Office of Management and Budget

The NRC submits its annual budget requests, including proposed personnel ceilings, to the Office of Management and Budget for approval.

U.S. Department of State

By law, the NRC must license the export and import of nuclear equipment and material. On significant applications, the Commission requests the U.S. Department of State to provide it with Executive Branch views on whether the license should be issued.

The NRC also works with the U.S. Department of State in such matters as negotiating international agreements in the nuclear field; dealing with such international organizations as the IAEA and other organizations of the United Nations, and the Nuclear Energy Agency of the Organization for Economic Cooperation and Development, developing policy on nuclear issues that are under the NRC's purview; and planning and coordinating programs of nuclear safety and safeguards assistance to other countries, such as the former Soviet Union and Central and Eastern Europe.

Environmental Protection Agency

The responsibilities of the NRC and the EPA cross or overlap in a number of areas in which the EPA issues generally applicable standards for activities that are also subject to NRC licensing. Examples include standards for high level waste repositories, decommissioning standards, and standards for public and worker protection.

Federal Emergency Management Agency (FEMA)

In 1979, after the accident at Three Mile Island, the President assigned FEMA the lead responsibility for offsite emergency planning and response at nuclear power plants. The NRC remained responsible for evaluating onsite planning, and for making the overall finding regarding whether a plant can operate "without undue risk to public health and safety." A 1980 Memorandum of Understanding between the two agencies, since updated, lays out the relationship between FEMA and the NRC, as it relates to emergency planning. Among other things, FEMA assists the NRC's licensing process by preparing reviews and evaluations, as well as presenting witnesses to testify at licensing hearings. FEMA also participates with the NRC in observing and evaluating emergency exercises at nuclear plants. FEMA's findings are not necessarily binding on the NRC, but they are presumed to be valid unless controverted by other more persuasive evidence.

Department of Labor

The NRC monitors discrimination actions filed with the Department of Labor under Section 211 of the Energy Reorganization Act and develops enforcement actions where there are properly supported findings of discrimination, either from the NRC's Office of Investigations or from the Department of Labor adjudications.

U.S. Department of Justice

NRC litigation almost always requires coordinating with the U.S. Department of Justice. Under the Administrative Orders Review Act (commonly called the "Hobbs Act"), the United States is a party to petitions for review challenging NRC licensing decisions or regulations. Department of Justice attorneys represent the United States. However, the Hobbs Act also provides for independent representation of the NRC by the agency's own attorneys. In practice this means that NRC attorneys, under the supervision of the NRC Solicitor, write the briefs in Hobbs Act cases and argue the cases in the courts of appeals. The Department of Justice typically joins in the NRC's briefs, except in the extremely rare circumstance in which the Department of Justice views an NRC position as inconsistent with general government interests.

In cases other than those involving the Hobbs Act (i.e., those cases (usually in district court) that do not involve NRC licensing or regulatory action, such as tort, subpoena enforcement, personnel, or Freedom of Information Act cases), Department of Justice attorneys normally take the lead role, with backup support from NRC attorneys.

The NRC's investigatory arms frequently work with the Department of Justice staff. The NRC has authority to revoke or suspend licenses, impose civil penalties, and take other civil actions for willful wrongdoing.

The Office of Investigations, which investigates allegations of wrongdoing by NRC's applicants and licensees, as well as by their contractors, normally deals with the General Litigation Section of the Criminal Division at the Department's headquarters and with U.S. Attorneys in the field.

Pursuant to the Inspector General Act, the Office of the Inspector General reports to the Department of Justice whenever it has reasonable grounds to believe that an NRC employee or contractor has violated Federal law. The Inspector General refers cases for review for possible criminal prosecution to the U.S. Attorney's Office for the area in which the potential violation occurred. When the Department of Justice desires support from the Office of the Inspector General for investigations or grand jury work, it makes the request directly to the Inspector General.

8.1.5.2 The States

At the NRC, relations with the States are primarily the responsibility of the Office of State and Tribal Programs.

As explained above, the Atomic Energy Act of 1954 confers the NRC with preemptive authority over health and safety regulation of nuclear energy. (The Clean Air Act subsequently gave the EPA some authority over air emissions of radionuclides, and that authority was reconfirmed by Congress in 1990.) As a result, the general rule is that nuclear safety, like airline safety, is the exclusive province of the Federal Government and cannot be regulated by the States. A State law that attempted to set nuclear safety standards would thus be voided by the courts. However, the courts will not overturn a State law that regulates nuclear energy for purposes, such as economics, that are other than health and safety, unless it conflicts with an NRC requirement. Similarly, the courts will not ordinarily question a State's declared purpose in enacting legislation.

The Atomic Energy Act of 1954 did not entirely exclude States from nuclear regulation. Section 274 of the Act created the Agreement State Program, under which the NRC may delegate its authority over most nuclear materials to States that are willing and are technically able to assume that responsibility. The NRC may not relinquish authority over such facilities as reactors, fuel reprocessing and enrichment plants, imports and exports, critical mass quantities of special nuclear materials, high-level-waste disposal, or certain other excepted areas. When a State applies to become an Agreement State, the NRC reviews its regulatory program for adequacy and compatibility with the NRC's own program.

There are now over 30 Agreement States. Despite the fact that Agreement States receive no Federal funding to support their regulatory programs, only one State (Idaho) has ever relinquished Agreement State status after once assuming it.

Some States have shown a desire to participate in matters relating to nuclear power plants. In response, the NRC issued a policy statement in February 1989 declaring its intent to cooperate with States in the area of nuclear safety by keeping States informed of matters of interest to them, and considering State proposals under which State officials would participate in NRC inspection activities, pursuant to a Memorandum of Understanding between the State and the NRC. The policy statement makes clear that States must channel their contacts with the NRC through a single State Liaison Officer, appointed by the Governor. States are authorized only to observe and assist in NRC inspections, not to conduct their own independent health and safety inspections.

The Nuclear Waste Policy Act of 1982 also affords a major role to affected States and Native American tribes in decisions on siting high-level waste repositories and monitored retrievable storage facilities.

8.1.5.3 Congress

This section explains the relationship of the NRC to the U.S. Congress. Components of Congress discussed are the NRC's oversight committees in the Senate and House and subcommittees with jurisdiction over aspects of the NRC's activities.

Senate Oversight

Unlike most Federal agencies, the relations of the NRC with its oversight committees are (in part) governed by statute. The NRC is required to keep these committees "fully and currently informed" of matters in the NRC's jurisdiction.

In the Senate, the Committee on Environment and Public Works exercises jurisdiction over domestic nuclear regulatory activities. Within the Committee, the Subcommittee on Transportation, Infrastructure, and Nuclear Safety has responsibility for legislation and oversight of the NRC. It considers nominations of Commissioners and the Inspector General.

The Senate Energy and Natural Resources Committee shares jurisdiction over nuclear waste issues with the Environment and Public Works Committee.

House Oversight

In the U.S. House of Representatives, jurisdiction over domestic nuclear regulatory activities resides in the Committee on Energy and Commerce. Within the Committee, the Subcommittee on Energy and Power has responsibility for regulation and oversight of the NRC.

Other Relevant Committees

In addition to these subcommittees, the House and Senate Appropriations Subcommittees on Energy and Water Development play a key role in approving the Commission's annual budget. A number of other Congressional subcommittees on appropriations, international affairs, research, and general government operations have jurisdiction over some aspect of NRC activities.

8.2 Separation of Functions of the Regulatory Body from Those of Bodies Promoting Nuclear Energy

This section explains how the functions of the NRC are separate from those of the U.S. Department of Energy (DOE). The DOE, in addition to its responsibilities over nuclear weapons, has the responsibility of promoting the peaceful use of nuclear power through research, grant, and demonstration projects.

The NRC is required under the Energy Reorganization Act to license the disposal activities for all high-level radioactive defense wastes produced at DOE facilities. For example, the Commission is required to license any DOE facility that is primarily used for the receipt, storage, or disposal of high-level radioactive waste that is generated from NRC-licensed activities; any DOE facility that is authorized for long-term storage or disposal of high-level radioactive waste from defense activities; any DOE facility that is authorized for disposal of certain commercial low-level waste (so-called "greater than Class C waste"); and any Monitored Retrievable Storage Facility that is designated for the storage of spent fuel. This would include the high-level-waste repository that is currently under consideration at Yucca Mountain in Nevada.

The NRC will also license the proposed DOE mixed-oxide fuel fabrication facility which is part of the U.S. program to eliminate weapons grade plutonium.

The NRC does not have regulatory responsibility over research and weapons-related nuclear facilities of the DOE, such as Rocky Flats in Colorado, Fernald in Ohio, and the Hanford Reservation in Washington. The NRC also does not have responsibility to regulate the disposal of the DOE's low-level radioactive waste (other than commercial "greater than Class C").

The NRC and the DOE share mutual assistance agreements for responding to radiological emergencies through the Federal Response Plan and the Federal Radiological Emergency Response Plan coordinated by FEMA.

ARTICLE 9. RESPONSIBILITY OF THE LICENSE HOLDER

Each Contracting Party shall ensure that prime responsibility for the safety of a nuclear installation rests with the holder of the relevant license and shall take the appropriate steps to ensure that each such license holder meets its responsibility.

This section explains how the NRC ensures that each licensee meets its primary responsibility for safety. The NRC ensures the safety of nuclear installations through its licensing process, discussed in Articles 18 and 19, its Reactor Oversight Process, discussed in Article 6, and its Enforcement Program, discussed below.

9.1 Introduction

The basis for the NRC's regulatory programs continues to be that the safety of commercial nuclear power reactor operations is the responsibility of NRC licensees. The responsibility of the NRC is regulatory oversight of licensee activities to ensure safety is maintained. The NRC comprehensively reviews the safety of a reactor design and the capability of an applicant to design, construct, and operate a facility. If an applicant satisfies the requirements of the *Code of Federal Regulations*, the NRC then issues a license to operate the facility. Such licenses specify the terms and conditions of operation, to which a licensee must conform. Failure to conform would subject the licensee to enforcement action, which can include modifying, suspending, or revoking the license. The NRC can also order particular corrective actions or issue civil penalties. These enforcement mechanisms are discussed in greater detail below.

9.2 NRC Enforcement Program

This section explains the NRC enforcement program, and provides experience and examples.

9.2.1 Description of Program

As discussed in Article 7, the NRC has enforcement powers. As discussed in Article 6 (section 6.2.2.5), the enforcement process has been integrated into the revised reactor oversight process. The NRC uses enforcement as a deterrent to emphasize the importance of compliance with regulatory requirements, and to encourage prompt identification and prompt, comprehensive correction of violations.

The NRC identifies violations through inspections and investigations. All violations are subject to civil enforcement action and may also be subject to criminal prosecution. After the NRC identifies an apparent violation, the agency assesses it in accordance with the Commission's Enforcement Policy. The Policy is published as NUREG-1600, "General Statement of Policy and Procedure for NRC Enforcement Actions," to foster its widespread dissemination.   Periodically revised, NUREG-1600 is a living policy statement; its revisions are noticed in the *Federal Register*. The NRC's Office of Enforcement maintains the current policy statement on its Website at *www.nrc.gov/OE*. Because the Enforcement Policy is not a regulation, the Commission may deviate from it as appropriate, depending on the circumstances of a particular case.

The NRC has three primary enforcement sanctions available: Notices of Violation, civil penalties, and Orders.[2] A Notice of Violation identifies a requirement and how it was violated, formalizes a violation pursuant to 10 CFR 2.201,"Notice of Violation," requires corrective action, and normally requires a written response. A civil penalty is a monetary fine issued under authority of Section 234 of the Atomic Energy Act or Section 206 of the Energy Reorganization Act. Section 234 of the Atomic Energy Act provides for penalties of up to $100,000 per violation per day; however, that amount is adjusted every four years by the Debt Collection Improvement Act of 1996, and is currently $120,000. The Commission's Order issuing authority under Section 161 of the Atomic Energy Act is broad, and extends to any area of licensed activity that affects public health and safety. Orders modify, suspend, or revoke licenses, or require specific actions by licensees or persons. Notices of Violations and civil penalties are issued on the basis of violations. Orders may be issued for violations, or, in the absence of a violation, because of a public health or safety issue.

The NRC first assesses the significance of a violation by considering: (1) actual safety consequences; (2) potential safety consequences; (3) potential for affecting the NRC's ability to perform its regulatory function; and (4) any willful aspects of the violation. Violations are either (1) assigned a severity level, ranging from Severity Level IV for those of more than minor concern to Severity Level I for the most significant, or, (2) associated with findings assessed through the significance determination process of the Revised Reactor Oversight Process that are assigned a color of green, white, yellow, or red according to increasing risk significance. The significance determination process of the reactor oversight process is described in Article 6.

The way in which the NRC dispositions a violation is intended to reflect the seriousness of the violation and the associated circumstances. Most of the violations identified in the nuclear industry are of low risk significance. Provided certain criteria in Section VI.A of the Enforcement Policy are met, the NRC will normally disposition Severity Level IV violations and violations associated with green findings as Non-Cited Violations. The NRC will document Non-Cited Violations in inspection reports (or inspection records for some materials licensees) to establish public records of the violations, but will not cite them in Notices of Violations, which normally require written responses from licensees. Dispositioning violations in this way does not reduce the NRC's emphasis on compliance with requirements nor the importance of maintaining safety. Licensees are still responsible for maintaining safety and compliance and must take steps to address corrective actions for violations. Licensees must correct even minor violations. However, given their limited risk significance, minor violations are not subject to enforcement action and are not normally described in inspection reports. This treatment of violations that have low risk significance is consistent with the agency's performance goals.

The NRC may hold a predecisional enforcement conference or a regulatory conference with a licensee before making an enforcement decision, if the NRC concludes that a conference is necessary or the licensee requests it, and escalated enforcement actions appear to be

---

[2]The NRC also uses administrative actions, such as Notices of Deviation, Notices of Nonconformance, Confirmatory Action Letters, and Demands for Information to supplement its enforcement program.

warranted. (An escalated enforcement action is defined as an action involving Severity Level I, II, or III violations; violations associated with white, yellow, or red findings; civil penalties; or orders.) The conference is intended to reach a common understanding about facts, root causes, and missed opportunities associated with the apparent violations, corrective action taken or planned, or the significance of issues and the need for lasting comprehensive corrective action. Such an understanding helps the NRC determine the proper enforcement action. The decision to hold a conference does not mean that the agency has determined that a violation has occurred, or that it will take enforcement action. In accordance with the Enforcement Policy, conferences are normally open to public observation. If the NRC concludes that a conference is not necessary, it may give a licensee an opportunity to respond to the apparent violations before making an enforcement decision.

The NRC normally assesses civil penalties for Severity Level I and II violations and knowing and conscious violations of the reporting requirements of Section 206 of the Energy Reorganization Act. It considers civil penalties for Severity Level III violations. Although the NRC will not normally use civil penalties for violations associated with the Reactor Oversight Process, it will consider civil penalties (and the use of severity levels) for issues that are willful, that have the potential for affecting the regulatory process, or that have actual consequences.

The NRC imposes different levels of civil penalties on the basis of a combination of the type of licensed activity, the type of licensee, the severity level of the violation, and (1) whether the licensee has had any previous escalated enforcement action (regardless of the activity area) during the past 2 years or past two inspections, whichever is longer; (2) whether the licensee should be given credit for actions related to identification; (3) whether the licensee's corrective actions are prompt and comprehensive; and (4) whether, in view of all the circumstances, the matter in question requires the exercise of discretion. Although each of these decisional points may have several associated considerations for any given case, the outcome of the assessment process for each violation or problem, absent the exercise of discretion, is limited to one of the following three results: no civil penalty, a base civil penalty, or twice the base civil penalty.

If the NRC proposes a civil penalty, it issues a written Notice of Violation and Proposed Imposition of Civil Penalty. The licensee has 30 days to respond in writing, by either paying the penalty or contesting it. The NRC considers the response, and if the penalty is contested, may either mitigate the penalty or impose it by order. Thereafter, the licensee may pay the civil penalty or request a hearing.

In addition to proposing civil penalties, the NRC may use Orders to modify, suspend, or revoke licenses. Orders may require additional corrective actions, such as removing specified individuals from licensed activities or requiring additional controls or outside audits. Persons who are adversely affected by Orders that modify, suspend, or revoke a license, or that take other action, may request a hearing.

The NRC issues a press release with a proposed civil penalty or Order. All Orders are published in the *Federal Register*.

9.2.2 Experience and Examples

Starting in 1998, the enforcement program began to undergo significant change. In the following years, the enforcement program has continued to change as an integral element of the agency's oversight process together with meeting the agency's revised reactor performance goals of maintaining safety, reducing unnecessary burden, making NRC activities and decisions more effective, efficient, and realistic, and increasing public confidence.

Changes in the enforcement program have stemmed from changes in inspection initiatives (e.g., revised reactor oversight process) as well as from changes within the enforcement program itself. Examples of changes within the program are eliminating the practice of aggregating multiple low significance violations into escalated enforcement and establishing a management-level review group to evaluate the processes for handling discrimination issues. The enforcement program will continue to change as a result of new initiatives in the inspection programs and new initiatives from within the enforcement program itself. This process reflects the NRC's extensive efforts to address industry and other stakeholder concerns and demonstrates the agency's commitment to more risk-informed, performance-based regulatory and enforcement programs.

During fiscal year 2000 (October 1, 1999, through September 30, 2000), the NRC issued a variety of enforcement actions to operating power reactors. Specifically these actions were 18 escalated Notices of Violation without civil penalty; 3 civil penalties, one Order, and 2 Orders imposing civil penalties. These statistics reflect a reduction in the number of escalated enforcement actions issued during a fiscal year. This reduction results from the agency's efforts to establish an enforcement program that strives to be more risk-informed and performance-based.

To provide accurate and timely information to all interested stakeholders and enhance the public's understanding of the enforcement program, the NRC's Office of Enforcement electronically publishes enforcement information on its Website. This information includes copies of significant enforcement actions issued to reactor and materials licensees and individuals since 1996. The following examples of enforcement actions illustrate the various ways that operating reactor violations can be dispositioned in the NRC's enforcement program. Colors of findings refer to the color coding of the Significance Determination Process of the Revised Reactor Oversight Process.

Niagara Mohawk Power Corporation (Nine Mile Point Nuclear Station) EA 01-011

On May 2, 2001, a Notice of Violation was issued for a Severity Level III violation involving the deliberate failure of an NRC-licensed chief shift operator to provide complete and accurate information on health history forms that were required for the Fitness-For-Duty regulations.

Nuclear Management Company, LLC (Kewanee Nuclear Power Plant) EA 00-214

On February 28, 2001, a Notice of Violation was issued for a violation associated with a white finding involving the emergency response staffing drills. The violation was based on the fact that timely augmentation of response capabilities was not available and that the licensee failed to correct deficiencies that were identified as a result of several monthly drill failures.

Nuclear Management Company, LLC (Prairie Island Nuclear Power Plant) EA 00-282

On February 20, 2001, a Notice of Violation was issued for a violation associated with a white finding involving the potential inability of the deep cooling water (service water) pumps to perform their intended safety function. The violation was based on the licensee's failure to ensure that design control measures would verify the adequacy of the design and would assure that appropriate quality standards were specified.

Union Electric Company (Callaway) EA 00-208

On January 9, 2001, a Notice of Violation was issued for a violation associated with three white findings involving performance deficiencies in the licensee's procedures and engineering controls designed to achieve occupational doses that are as low as is reasonably achievable (ALARA).

Consolidated Edison Company of New York (Indian Point 2) EA 00-179

On November 20, 2000, a Notice of Violation was issued for a violation associated with a red finding. The violation involved the licensee's failure to identify and correct a significant condition adverse to quality involving the presence of flaws from primary water stress corrosion cracking in the steam generator tubes, despite opportunities to act during the 1997 refueling outage. As a result, one of the tubes failed on February 15, 2000, when the reactor was at 100% power.

Tennessee Valley Authority (Browns Ferry Nuclear Plant) EA 00-163

On October 27, 2000, a Notice of Violation was issued for a Severity Level III violation involving the willful failure to perform required evaluations for out-of-tolerance measuring and test equipment.

Vermont Yankee Nuclear Power Corporation (Vermont Yankee) EA 00-165

On September 18, 2000, a Notice of Violation was issued for a Severity Level III violation. The action was based on a former mechanical maintenance manager deliberately causing a violation of the procedure implementing the requirement to control contracted services during the 1998 refueling outage.

Entergy Operations, Inc. (Waterford 3) EA 00-093

On August 4, 2000, an immediately effective Confirmatory Order was issued to confirm commitments made by the licensee to perform corrective actions for the physical security program at the Waterford 3 facility.

Tennessee Valley Authority (Sequoyah Nuclear Plant) EA 99-234

On February 7, 2000, a Notice of Violation and Proposed Imposition of Civil Penalty in the amount of $110,000 was issued for a Severity Level II violation involving employment discrimination against a former corporate employee for engaging in protected activities.

ARTICLE 10. PRIORITY TO SAFETY

Each Contracting Party shall take the appropriate steps to ensure that all organizations engaged in activities directly related to nuclear installations shall establish policies that give due priority to nuclear safety.

This section focuses on probabilistic risk assessment (PRA) as a major element of a policy giving due priority to safety. Specifically, this section covers the policy, safety goals and objectives, and applications of PRA. The applications discussed are (1) the use of safety goals to resolve severe accident issues and evaluate new and existing regulatory requirements and programs, (2) the implementation plan for risk-informed regulation, (3) activities that improve data and methods of risk analysis, (4) industry activities and pilot PRA applications, and (5) activities that apply risk assessment to plant-specific changes to the licensing basis.

Other articles, for example, Articles 6, 14, 18, and 19, also discuss activities undertaken to achieve nuclear safety at nuclear installations.

In particular, see the discussion of the Revised Reactor Oversight Process in Article 6.

10.1 Background

The U.S. has made much progress in developing and using the results of PRAs for all operating reactor facilities, and the NRC has developed extensive guidance regarding the role that PRA is to play in regulatory programs in the U.S. Specifically, the NRC uses insights derived from PRA, together with safety goals, to prioritize resources, establish regulatory coherence, and develop policies that give due priority to nuclear safety. The use of risk assessment and safety goals in establishing safety priority in nuclear regulatory programs has been especially timely in light of Executive Order 12866, "Regulatory Planning and Review." This Order requires regulatory agencies to consider the degree and nature of risks posed in setting their regulatory priorities, as well as the costs and benefits of intended regulation. NRC's large investment and substantial experience in PRA ensures that it has developed a systematic approach to give due priority to safety.

The NRC believes that a PRA for a plant can yield information about plant safety that no other methodology can produce. The NRC has extensively applied information gained from PRA to complement other engineering analyses in improving issue-specific safety regulation, and in changing the current licensing bases for individual plants. The move toward risk-informing the current regulations and processes marks perhaps the most significant changes taking place at the NRC. An example of risk-informing the current regulations, the rulemaking plan for modifying the scope of the 'special treatment' regulations in 10 CFR Part 50, is close to completion. This plan will allow for an alternative regulatory framework that will enable licensees to use a risk-informed approach to categorize structures, systems, and components, and their associated design and protection, according to their safety significance.

The NRC is also proceeding with a program to develop changes to the specific technical requirements in the body of 10 CFR Part 50, including the general design criteria. This program has provided the Commission with a framework for risk-informing deterministic requirements.

The staff has used this framework to make recommendations to the Commission about the technical feasibility of risk-informing combustible gas control and acceptance requirements for the emergency core cooling system. The NRC is proceeding with rulemaking for combustible gas control expeditiously and in close cooperation with its stakeholders.

10.2 PRA Policy

Three policy statements form the basis for the NRC's current treatment of PRA and the related regulatory safety goals and objectives. The first is the "Policy Statement on Severe Reactor Accidents Regarding Future Designs and Existing Plants," issued August 8, 1985; the second is the "Safety Goals for the Operations of Nuclear Power Plants Policy Statement," issued August 21, 1986 and the third is the "Policy Statement on Use of PRA Methods in Nuclear Activities," issued August 16, 1995.

The chief aim of the policy statement on severe accidents was to ensure that a licensee takes all reasonable steps to prevent (i.e., reduce the likelihood of) a severe accident substantially damaging the reactor core, and to mitigate the consequences of such an accident should one occur. The focus on severe accident issues was prompted by the NRC's judgment that such accidents, beyond the traditional design-basis events, carry the potential for a major risk of radioactive releases from nuclear power to the public.

The aim of the safety goal policy statement was to establish goals that broadly define an acceptable level of radiological risk that might be imposed on the public as a result of nuclear power plant operation. Radiological risk means the risk associated with the release of radioactive material from the reactor to the environment from normal operations as well as from accidents.

The safety goals, which are qualitative, are as follows:

(1) Individual members of the public should be provided a level of protection from the consequences of nuclear power plant operation such that individuals bear no significant additional risk to life and health.

(2) Societal risks to life and health from nuclear power plant operation should be comparable to or less than the risk of generating electricity by viable competing technologies, and should not be a significant addition to other societal risks.

To define "risk to life and health," the NRC approved the following quantitative health objectives:

(1) The risk to an average individual in the vicinity of a nuclear power plant of prompt fatalities that might result from reactor accidents should not exceed one-tenth of one percent (0.1 percent) of the sum of prompt fatality risks resulting from other accidents to which members of the U.S. population are generally exposed.

> (The average individual in the vicinity of the plant is defined as the average individual biologically, in terms of age and other risk factors, who resides within 1 mile from the plant site boundary. This means that the risk to the average individual is found by

accumulating the individual risks and dividing them by the number of individuals residing in the vicinity of the plant.

(2) The risk of fatalities from cancer to the population in the area near a nuclear power plant that might result from nuclear power plant operation should not exceed one-tenth of one percent (0.1 percent) of the sum of cancer fatality risks resulting from all other causes.

The population considered "near" a nuclear power plant is taken as the population within a 10-mile radius of the plant site.

In its policy statement on the "Use of Probabilistic Risk Assessment Methods in Nuclear Regulatory Activities," the NRC established an overall policy to ensure that PRA would be implemented in its potential applications in a manner that promotes regulatory stability and enhances safety. The policy was intended to encourage both the NRC and the industry to use PRA and expand its scope of applications in regulatory matters to the extent supported by the state of the art. In its policy statement, the NRC also confirmed its intention to use its safety goals and subsidiary objectives, properly considering the uncertainties in assessments, to determine the need for proposing and backfitting new generic requirements.

10.3 Applications of PRA

This section discusses applications of PRA. The NRC applies PRA to resolve severe accident issues, evaluate new and existing requirements and programs, implement risk-informed regulation, and improve data and methods of risk analysis. The NRC also engages in cooperative activities with industry (such as the Graded Quality Assurance Pilot Program and risk-informed inservice testing) and in activities that assess risk in determining plant-specific changes to the licensing basis.

10.3.1 Severe Accident Issues

The main focus of the severe accident policy statement was on the criteria and procedures to be used to certify new designs for nuclear power plants. The NRC expected new plants to achieve a higher standard of safety performance in severe accidents than plants of earlier designs. Demonstrating its commitment to the NRC's severe accident policy, the industry established a goal of designing future reactors to a core damage frequency of less than 1×10^{-5} per year of reactor operation. The "Probabilistic Safety Assessment Applications Guide" for advanced light-water reactors, published in August 1995, by the Electric Power Research Institute, describes a method to meet this goal.

As for addressing the risk of severe accidents at existing plants, the NRC established the "Integration Plan for Closure of Severe Accident Issues" (SECY 88-147), confirming the view that it saw no need for immediate regulatory action. Elements of the plan involved performing an individual plant examination to identify significant plant-specific vulnerabilities for internally and externally initiated events. The plan also involved assessing generic improvements to containment performance, the result of which led to licensees' installing hardened containment vents in Mark I containments for boiling water reactors and recommending additional containment improvements for future consideration. Other elements of the integration plan involved improving plant operation, establishing a research program into severe accidents, and establishing programs for accident management.

As part of the Integration Plan, the Individual Plant Examination Program examines operating plants for vulnerabilities to severe accidents attributable to internally initiated *events within the plant* during full power operation. By contrast, the Individual Plant Examination of External Events program examines vulnerabilities to severe accidents caused by *external events*, such as earthquakes, fires, and high wind. These programs have resulted in licensees using risk assessment methods to identify plant vulnerabilities that need attention. The NRC reviewed all of the plant-specific risk studies that have emerged from the individual plant examination programs and expedited its review of some programs that revealed high estimated core damage frequencies or previously unknown vulnerabilities. The NRC's review of these programs for all of the existing plants showed core damage frequencies that are generally in the range of 1×10^{-4} to 1×10^{-6} per reactor year, and revealed that loss-of-offsite power and station blackout are the significant contributors to core damage frequency. As a result of its review, the NRC published NUREG/CR-1560, "Individual Plant Examination Program: Perspectives on Reactor Safety and Plant Performance," in December 1997. The NRC also recently issued for public comment, a draft report, NUREG-1742, "Perspectives Gained from the Individual Plant Examination of External Events Program."

The NRC referred to such documents, as NUREG/BR-0058, Revision 3, "Regulatory Analysis Guidelines of the U.S. Nuclear Regulatory Commission," issued in July 2000, in determining when to take prompt action as a result of its review of the individual plant examinations. In general, for analyses that showed a core damage frequency of 1×10^{-3} per reactor-year or greater, the NRC considered prompt action. For core damage frequencies in the range of 1×10^{-6} to 1×10^{-7} or less, the agency took no action. For core damage frequencies of 1×10^{-4} to 1×10^{-6}, the NRC considered the probability of early containment failure in deciding whether to pursue cost-beneficial enhancements.

10.3.2   Using the Safety Goals to Evaluate New and Existing Regulatory Requirements and Programs

Underlying the safety goals is the premise that the current regulatory practice of requiring compliance with the NRC's regulations ensures the basic statutory standard of adequate protection. The NRC believes, however, that the current practices can be improved to better test the adequacy and necessity of current requirements, as well as the possible need for additional requirements. The NRC sees the safety goals policy as a vehicle for achieving these objectives. As a result, the safety goals have become the bases, in part, for evaluating the need for new and revised regulations and regulatory practices. In fact, the NRC has applied this ongoing evaluation and refinement to simplify the implementation of the safety goals themselves. In doing so, the NRC recognized that, although the safety goals are quite straightforward, they are somewhat difficult to implement. As a result, the NRC established a subsidiary objective of a core damage frequency of 1×10^{-4} per reactor-year. In addition, the NRC approved a conditional containment failure probability of 0.1 (one-tenth) for evolutionary light water reactor designs. These values have evolved into the "benchmark" values of 1×10^{-4} for core damage frequency and 1×10^{-5} for large early release frequency, as discussed in Regulatory Guide 1.174, " An Approach for Using Probabilistic Risk Assessment in Risk-Informed Decisions on

Plant-Specific Changes to the Licensing Basis," for use in risk-informed regulatory decision making.

Over the past several years, the agency has used these subsidiary objectives in developing new regulations. For example, it developed new regulations on anticipated transients without scram, station blackout, and pressurized thermal shock, in part, using the estimated changes to the collective core damage frequency provided by the rules, and by applying the subsidiary objectives. Other policies and regulations that have been based, in part, on PRA methods and insights are the "Final Policy Statement on Technical Specification Improvement for Nuclear Power Reactors," the alternative to 10 CFR Part 50 Appendix J, "Containment Leakage," and the reactor siting criteria in the 1996 revision to 10 CFR Part 100, "Reactor Site Criteria." (As explained in Article 17, these siting criteria invoked a probabilistic approach to estimate the ground motion for a safe shutdown earthquake for a nuclear reactor site, instead of the purely deterministic method specified in Appendix A, "Seismic and Geologic Siting Criteria for Nuclear Power Plants," to 10 CFR Part 100.)

The NRC is committed to ensuring that future regulatory initiatives will conform to the quantitative objectives of the safety goals. Additionally, in reviewing the need to apply new requirements to existing plants, the NRC will consider the safety goals when evaluating the potential benefit expected from the new requirements. The process is discussed in the NRC's backfit rule, 10 CFR 50.109, "Backfitting."

The NRC also applies PRA and safety goal principles to enhance existing programs. For example, the NRC uses PRA techniques in the Accident Sequence Precursor Program to evaluate conditional core damage frequencies. It applies the safety goals when setting priorities for resolving generic issues under the Program for Resolving Generic Issues.

10.3.3 Risk-Informed Regulation Implementation Plan

The NRC's "Risk-Informed Regulation Implementation Plan" describes how the agency will further incorporate PRA into its safety policies and regulatory activities. The staff presented the plan for Commission information in SECY-00-0213, dated October 26, 2000, and will update it in the future after a period for public comment. Activities that were previously included within the superseded "PRA Implementation Plan" are now covered in this new plan, which has (1) a statement of objectives linked to the NRC's Strategic Plan, (2) a set of criteria and a process for deciding which activities are amenable to risk treatment, (3) guidelines for risk-informed activities, (4) a summary of activities that are currently planned to implement the risk-informed strategies, (5) a description of planned communications, and (6) a description of training plans. The activities fall into three major arenas, encompassing reactor safety, nuclear materials, and waste (including spent fuel).

The activities described in the "Risk-Informed Regulation Implementation Plan" include initiatives that are currently underway, such as the revised Reactor Oversight Process, as well as developmental efforts. These activities apply to a number of NRC interactions with the regulated industry, including developing guidance for NRC inspectors to use in focusing resources on risk-important equipment and reassessing plants with relatively high core damage frequencies for possible backfit. The NRC anticipates changes to regulations, guidance, inspection programs, and development of tools and data in the plan. The agency also recognizes that additional policy, technical, and legal issues may arise in later revisions.

10.3.4 Activities that Improve Data and Methods of Risk Analysis

The NRC's research activities consist of many issue-oriented projects, as well as more general work, such as developing and demonstrating risk analysis methods and risk-related training and guidance for the NRC staff. Progress has been made in (1) analysis of low-power and shutdown accident risks, (2) computer tools for running the Systems Analysis Programs for Hands-On Integrated Reliability Evaluation (SAPHIRE), (3) analysis of uncertainties of the effects of severe accidents on the population offsite, (4) human reliability analysis, (5) containment response to high-pressure melt ejection (direct containment heating), hydrogen combustion, core melt-concrete interactions, debris coolability, and fuel-coolant interactions, (6) source terms, (7) reactor vessel integrity under severe accident conditions, and (8) analytical codes for core melt progression.

SAPHIRE is an example of an improved risk analysis method that is used to perform PRAs. A work in progress, it is intended to permit an analyst to create, quantify, and evaluate the accident risk of nuclear power plants. For more information, see NUREG/CR-6116, "Systems Analysis Programs for Hands-on Integrated Reliability Evaluations (SAPHIRE) Version 5.0."

Another example of progress is an improved approach to human reliability analysis. This approach is intended to be fully integrated with the PRA methodology to improve the assessment of the human contribution to plant risk during low-power, shutdown and at-power operations. This approach is described in NUREG/CR-6093, "An Analysis of Operational Experience During Low-Power and Shutdown, and a Plan for Addressing Human Reliability Assessment Issues."

More recently, the staff has developed a process for human reliability analysis that addresses (1) how to identify and incorporate human failure events in the logic models that are used in PRAs, (2) information required for assigning probabilities to these failure events, (3) how to use this information to estimate the probabilities, and (4) how the probabilities incorporated into the PRA quantification process have been developed. The project is intended to show the usefulness and acceptability of the guidelines for implementing the methodology, using selected parts of a PRA.

The significance of cooperation to improve regulatory priority to safety is exemplified by the efforts of the NRC and stakeholders to establish a database concerning equipment reliability and availability to support the Maintenance Rule and other performance-based regulation. The NRC will continue to work with industry representatives and other stakeholders to identify areas of mutual interest for the use of PRA methods and insights to encourage the use of plant-specific failure data.

10.3.5 Industry Activities and Pilot PRA Applications

The NRC and industry representatives have cooperated in a number of activities and pilot programs to develop and apply risk-informed methodologies for specific regulatory applications. Lessons learned from these activities are used to enhance the effectiveness of developed guidance. The activities described in this section are graded quality assurance, inservice testing, inservice inspection, technical specification changes, and the developing of standards.

10.3.5.1 Graded Quality Assurance (QA) Pilot Program

The purpose of graded QA is to apply licensees' QA controls (such as reviews, inspections, and audits) in a manner that is consistent with plant equipment's importance to safety. Thus, graded QA allows both licensee and NRC staff to focus on more safety-significant equipment. Similarly, graded QA reduces the resources that must be allocated for QA activities for equipment of lesser safety significance. In general, existing licensee QA controls continue to apply to safety-significant equipment; less-rigorous licensee QA controls apply to equipment of lesser safety significance.

In 1997, the NRC and the industry developed a process to implement graded QA, similar to that implementing the Maintenance Rule. According to this process, a licensee's expert panel evaluates both PRA and deterministic evaluations to categorize plant equipment by safety significance. For guidance in implementing graded QA practices, the licensees for Palo Verde, Grand Gulf, and South Texas worked with the NRC to develop a regulatory guide (Regulatory Guide 1.176, "An Approach for Plant-Specific, Risk-Informed Decisionmaking", issued in August, 1998). In November 1997, the NRC approved the implementation of a graded QA program for the South Texas Project facility. That program is subject to periodic assessments of plant and industry information to adjust both quality controls and the categorization of safety significance. The QA Program is described in more detail in Article 13.

10.3.5.2 Risk-Informed Inservice Testing

In August 1998, the NRC issued Regulatory Guide 1.175, "An Approach for Plant-Specific, Risk-Informed Decisionmaking: Inservice Testing," which provides guidance regarding changes to the risk-informed inservice testing program. The agency subsequently completed a pilot application of risk-informed inservice testing in 1998, and has approved or is reviewing several other applications, generally of limited scope. The NRC also plans to review its experience and possibly revise the regulatory guide later in 2001.

10.3.5.3 Risk-Informed Inservice Inspection

In September 1998, the NRC issued Regulatory Guide 1.178, "An Approach for Plant-Specific Risk-Informed Decisionmaking: Inservice Inspection of Piping" and a companion Standard Review Plan chapter. The NRC has also approved two industry topical reports describing in detail two different methods to develop alternatives to the ASME Section XI Inservice Inspection program. The agency approved one method in December 1998, the other in October 1999. The NRC also approved three pilot applications during its review and approval of the topical reports. The agency subsequently approved 10 more applications, and is currently reviewing 16 others. Current applications include a new inservice inspection program for ASME Class 1 piping.

10.3.5.4 Risk-Informed Technical Specification Changes

In August 1998, the NRC issued Regulatory Guide 1.177, " An Approach for Plant-specific, Risk-Informed Decisionmaking: Technical Specifications," and a companion Standard Review Plan chapter. These documents provide guidance regarding risk-informed changes to plant technical specifications. The agency is using this regulatory guide as well as regulatory guide 1.174, " An Approach for Using Probabilistic Risk Assessment in Risk-Informed Decisions on Plant-Specific Changes to the Licensing Basis," to review and approve several applications.

10.3.5.5 Development of Standards

The NRC has been working with ASME to develop a national consensus standard for PRA covering internal initiating events. Staff members have been actively working with industry and ASME participants to resolve comments received on Revision 12 of the standard. A revision to the standard was released by ASME for public comment in June 2001, and final publication is expected in December 2001.

In parallel, the staff has been working with the American Nuclear Society to develop companion standards covering PRAs for external events, low power, and shutdown operations. The PRA standard for external events is currently issued for public comments, and a final revision was scheduled for August 2001. However, issuance of the final revision is being delayed until the ASME PRA Standard has been issued because the American Nuclear Society Standard relies heavily on references to the ASME Standard. Work on the low power and shutdown PRA Standard is progressing, and the final report is scheduled to be issued in June 2002.

In addition, the industry, represented by the Nuclear Energy Institute, issued NEI-00-02, "PRA Peer Review Process Guidelines," for a licensee to use in assessing whether its PRA is adequate to support different classes of risk informed application, and submitted it for NRC review. The NRC staff is preparing a document that indicates where the guidance in NEI-00-02 is insufficient to support no review of the PRA by NRC staff. The staff preliminarily concluded that, while the process can be of value in helping licensees understand the strengths and weaknesses of their PRAs, the process, as written, is not an adequate substitute for the expected ASME Standard.

10.3.6   Activities that Apply Risk Assessment to Plant-Specific Changes to the Licensing Basis

The risk-informed guidance documents described above are used by licensees to prepare proposals for plant-specific changes to their licensing bases. In addition, Regulatory Guide 1.174, "An Approach for Using Probabilistic Risk Assessment in Risk-Informed Decisions on Plant-Specific Changes to the Licensing Basis,"(issued in July 1998) and Standard Review Plan Chapter 19, "Use of Probabilistic Risk Assessment in Plant-Specific, Risk-Informed Decision Making: General Guidance," are notable for their guidance on using PRA to support licensees' requests for changes to a plant's licensing bases (such as license amendments and changes to technical specifications). Making use of the NRC's PRA policy statement, Regulatory Guide 1.174, provides guidance on evaluating the results of assessments for plant-specific changes. In this regulatory guide, the NRC has chosen a policy that permits, at most, only small increases in risk, and then only when sufficient margins and defense-in-depth are reasonably assured. The agency adopted this policy because of uncertainties in analysis and because, despite the maturity of the nuclear power industry, safety issues continue to emerge in design, construction, and operation. These factors suggest that nuclear power reactors should operate with a prudent safety margin above the level that provides adequate protection. The NRC has been approving many requests for license amendments on the basis of the guidance of Regulatory Guide 1.174. For details concerning this guidance, see the NRC Website: *http://www.nrc.gov.*

ARTICLE 11. FINANCIAL AND HUMAN RESOURCES

1. **Each Contracting Party shall take the appropriate steps to ensure that adequate financial resources are available to support the safety of each nuclear installation throughout its life.**

2. **Each Contracting Party shall take the appropriate steps to ensure that sufficient numbers of qualified staff with appropriate education, training, and retraining are available for all safety-related activities in or for each nuclear installation, throughout its life.**

This section explains the requirements regarding the financial resources that licensees must have to support the nuclear installation throughout its life, including the resources needed for financing safety improvements that are made during a plant's operation, decommissioning, and handling claims and damages associated with accidents. This section also explains the regulatory requirements for qualifying, training, and retraining personnel.

11.1 Financial Resources

Access to adequate funds for safe construction, operation, and decommissioning is necessary to ensure the protection of public health and safety. Although there does not appear to be a consistent relationship between a licensee's financial health and general indicators of operational safety, there is evidence that, at least for some plant owners and operators, financial pressures have limited the resources that are available to address corrective actions, plant improvements and upgrades, and other safety-related expenditures. Further, because a power reactor must operate to provide revenues for eventual plant decommissioning, any shutdown of a plant before its owner has accumulated sufficient funds for decommissioning could potentially hinder the safe, expeditious decommissioning of that plant.

Additionally, many States have initiated or completed actions to economically deregulate their nuclear power plants. Traditionally, nuclear power plant owners in many States have been large, vertically integrated companies with substantial generation, transmission, distribution, and other assets. In exchange for having exclusive franchises to provide electric power in defined geographical areas, plant owners have had the rates they charge to their customers regulated by State governmental bodies. This system of rate regulation has, in general, virtually ensured a source of funds for construction, operation, and decommissioning of nuclear power plants. Nonetheless, this model of rate regulation has been changing and, as discussed below, the U.S. has adjusted its processes.

The NRC distinguishes between financial qualifications for the construction, operation, and decommissioning of nuclear power plants, and has separate regulations and programs that apply to each. As part of its initial licensing reviews, the NRC must also determine whether activities that are conducted under a license will create or maintain a situation that is inconsistent with the antitrust laws of the U.S. However, because these reviews do not directly apply to the protection of public health and safety, they are not addressed further in this article. The NRC also implements programs to ensure that the public has financial protection for bodily injury and property damage losses in the event of an accident at a covered nuclear facility. Finally, the agency has implemented requirements to ensure that licensees have insurance to help pay onsite recovery costs resulting from accidents to provide funds for post-accident restart or decommissioning.

11.1.1   Financial Qualifications Program for Construction and Operations

This section explains the financial qualifications program for construction and operations. It covers the governing documents and process used to implement requirements. It explains NRC reviews for construction permits, operating licenses, combined licenses, post-operating non-transferred licenses, and transfers of licenses. It also describes experience and gives examples.

11.1.1.1 Governing Documents and Process

Section 182.a of the Atomic Energy Act of 1954, as amended (AEA), provides that "Each application for a license . . . shall specifically state such information as the Commission, by rule or regulation, may determine to be necessary to decide such of the technical and financial qualifications of the applicant . . . as the Commission may deem appropriate for the license." To implement this provision, the NRC has developed the following regulations and guidance:

(1) Regulations governing financial qualifications reviews of applications for licenses to construct or operate nuclear power plants are provided in Paragraph (f) of 10 CFR 50.33, "Contents of applications; general information."

(2) Guidance for financial qualifications reviews applying to construction permits is provided in Appendix C, "A Guide for the Financial Data and Related Information Required to Establish Financial Qualifications for Facility Construction Permits," to 10 CFR Part 50.

(3) Financial qualifications applying to license transfers are addressed in 10 CFR 50.80, "Transfers of Licenses."

(4) The overall process for NRC review of applicants' and licensees' financial qualifications for nuclear power plant construction and operation is described in the "Standard Review Plan on Power Reactor Licensee Financial Qualifications and Decommissioning Funding Assurance," NUREG-1577, Rev. 1.

Construction Permit Reviews

10 CFR 50.33(f)(1) requires applicants for construction permits to submit information that "demonstrates that the applicant possesses or has reasonable assurance of obtaining the funds necessary to cover estimated construction costs and related fuel cycle costs." Appendix C to 10 CFR Part 50 provides more specific directions for evaluating the financial qualifications of applicants. In accordance with Appendix C, applicants should submit at least three types of information:

(1) an estimate of construction costs, including plant costs ascribable to the nuclear plant itself, general and overhead plant costs, including transmission and distribution costs ascribable to the plant; and nuclear fuel costs for the first core load

(2) the source(s) of construction funds, including a financial plan describing internal and external sources of funds

(3) the latest published annual financial reports, together with any current interim financial statements that are pertinent, including income, balance sheet, and cash flow statements

In addition, newly established organizations must provide information showing the following details:

(1) the legal and financial relationships that they have or propose to have with their stockholders, corporate affiliates, and others (such as financial institutions) upon which they are relying for financial assistance; and, if the sources of funds upon which applicants intend to rely include parent companies or other corporate affiliates, applicants should include information to support the financial capability of each such company or affiliate to meet its commitments to the applicants

(2) operating, generating, or service company subsidiaries

(3) any other information that may be considered necessary by the Commission to enable it to determine the applicant's financial qualifications

(4) the applicant's statements of assets, liabilities, and capital structure, as of the date of the application

Operating License Reviews

An "electric utility" as defined in 10 CFR 50.2, "Definitions," is "any entity that generates or distributes electricity and which recovers the cost of this electricity, either directly or indirectly, through rates established by the entity itself or by a separate regulatory authority." Electric utilities are exempt under 10 CFR 50.33(f) from reviews of financial qualifications of applications for operating licenses. The reason for this exemption is that traditional cost-of-service rate regulation, as it has existed in the U.S., has virtually ensured that ratepayers provide a source of funds for the safe operation of nuclear power plants. Applicants for operating licenses that are not "electric utilities" are required under 10 CFR 50.33(f)(2) to submit information that demonstrates that they possess or have reasonable assurance of obtaining the necessary funds to cover estimated operating costs for the period of the license. Non-electric-utility applicants for operating licenses are also required to submit estimates for the total annual operating costs for each of the first 5 years of operation of their facilities, and must also indicate the sources of funds to cover operating costs.

Combined License Application Reviews

As authorized in 10 CFR Part 52, "Early Site Permits; Standard Design Certifications; and Combined Licenses for Nuclear Power Plants," applicants may apply for a combined construction permit and operating license. In accordance with 10 CFR 52.77, "Contents of applications; technical information," all such applications must contain all of the information required under 10 CFR 50.33, including information regarding financial qualifications. The review procedures described above are used to review any combined license applications that the NRC receives.

Post-Operating License Non-Transfer Reviews

The NRC does not systematically review the financial qualifications of power reactor licensees once it has issued an operating license, other than for license transfers as described below. However, as provided in 10 CFR 50.33(f)(4), the NRC can seek additional information on licensees' financial qualifications if the agency considers such information appropriate.

Reviews of License Transfers

NRC regulations in 10 CFR 50.80 require agency review and approval of transfers of operating licenses, including licenses for nuclear power plants that are owned or operated by "electric utilities." The NRC performs these reviews to determine whether a proposed transferee is technically and financially qualified to hold the license. The NRC evaluates the financial qualifications of the proposed transferees through the following activities:

(1) Determine whether the transferee will remain an "electric utility" after the transfer.

(2) If the transferee is not an "electric utility," review the recent financial performance of the proposed transferee or, if the proposed transferee is a new entity, such as an operating, generating, or service company subsidiary of an existing licensee, review the participation agreement with its owners or other responsible parties.

(3) Identify all parent companies relevant to the transfer that are not licensed by the NRC or that did not undergo an NRC transfer review.

11.1.1.2 Experience and Examples

The NRC has not received a construction permit application for a nuclear power plant for more than 20 years. Thus, the agency has not performed any financial qualifications reviews of applications for construction permits since the 1970s. The last power reactor unit to receive an operating license was Watts Bar Unit 1, in 1996. Because Watts Bar is owned and operated by the Tennessee Valley Authority, which meets the NRC's definition of an "electric utility," the facility is exempt from an operating license financial qualifications review.

However, the NRC has conducted several post-licensing reviews under the license transfer provisions of 10 CFR 50.80. To foster free market competition, Federal and State governments have begun to economically deregulate the wholesale and retail sale of electricity under their jurisdictions; therefore, several NRC power reactor licensees have restructured themselves in various ways to meet increased competition. Since 1999, over ten plants have been sold. Additionally, power plant licensees have merged with other companies or have formed parent holding companies. The NRC has viewed each of these actions as a direct or indirect transfer of a license under the provisions of 10 CFR 50.80.

11.1.2 Financial Qualifications Program for Decommissioning

This section explains the financial qualifications program for decommissioning. First, it covers the governing documents and process to implement requirements. Next, it discusses decommissioning funding assurance for operating nuclear plants and for plants that have permanently shutdown. Finally, it discusses experience and gives examples.

11.1.2.1 Governing Documents and Process

Among other sections of the Atomic Energy Act, Section 182.a establishes the basis for the NRC's decommissioning funding assurance regulations and guidance, as follows:

(1) The NRC's regulations governing decommissioning funding assurance for plants under construction and operating nuclear power plants are contained in 10 CFR 50.75, "Reporting and Record Keeping for Decommissioning Planning."

(2) The NRC's regulations for nuclear power plants that have permanently ceased, or are about to permanently cease operations are provided in 10 CFR 50.82, "Termination of License."

(3) The overall process for NRC review of applicants and licensees with regard to decommissioning funding assurance is described in the "Standard Review Plan on Power Reactor Licensee Financial Qualifications and Decommissioning Funding Assurance," NUREG-1577, Rev. 1.

(4) Guidance on decommissioning funding assurance provisions and mechanisms is contained in "Assuring the Availability of Funds for Decommissioning Nuclear Reactors," Regulatory Guide 1.159. (On May 30, 2001, the NRC issued a revised draft of Regulatory Guide 1.159 as DG-1106 for public comment.)

Decommissioning Funding Assurance for Operating Nuclear Power Plants

Power reactor licensees were required to certify, by July 27, 1990, that they would have adequate funds to decommission each reactor unit by the time they plan to permanently shut down the unit. This certification was required to be predicated on an amount that is no less than, but may be greater than, the applicable generic decommissioning cost formulas in 10 CFR 50.75(c)(1) and (2). 10 CFR 50.75(c)(1) contains two formulas to determine the certification amounts (in 1986 dollars) for pressurized-water reactors and boiling-water reactors. The formulas include scaling factors to account for size differences in reactors. The decommissioning cost (in 1986 dollars) ranges from $85.6 million to $105 million for pressurized-water reactors, and from $114.8 million to $135 million for boiling-water reactors. In addition, 10 CFR 50.75(c)(2) contains a formula to determine the annual change in three primary decommissioning cost components, including labor, energy, and low-level-waste burial charges. Licensees are required to recalculate the formula amounts annually to account for changes (typically, inflation) in the three decommissioning cost factors during the previous year. Licensees are also required to report on the status of decommissioning funds, including current cost calculations, every 2 years.

The NRC's generic formulas in 10 CFR 50.75(c) contain only those decommissioning costs that apply to removing residual radioactivity sufficient to terminate the NRC license. Thus, costs of dismantling or demolishing nonradiological systems and structures, for example, are not included in the NRC's cost formulas. In addition, the costs of managing and storing spent fuel on site until transfer to the Department of Energy for permanent disposal are not included in the NRC's decommissioning cost formulas. Paragraph (bb) of 10 CFR 50.54 "Conditions of Licenses," requires licensees to collect those costs as operating costs years before permanently ceasing operations. Licensees who choose to use site-specific studies that include the costs of these activities should differentiate between the decommissioning costs and operating costs that the NRC requires. In addition, the NRC-required costs of the site-specific study must be equal to or greater than the generic formula amounts in Section 50.75(c).

The NRC's regulations in 10 CFR 50.75(f) require a licensee that intends to permanently shut down its power reactor within 5 years to develop a site-specific estimate. This estimate is intended to form a sufficient basis for collecting any remaining uncollected decommissioning funds by the time permanent shutdown occurs. The NRC selected 5 years as the period of time that is sufficiently close to the start of decommissioning so that performing site-specific decommissioning studies would not be overly burdensome. At the same time, 5 years would allow a licensee adequate time to collect any remaining decommissioning funds without causing undue financial stress in any single year. Thus, beginning at that 5-year milestone, the generic formulas would no longer normally be used.

In addition to establishing and updating the amount that is expected to be needed for decommissioning, licensees are required to use funding mechanisms that, in the NRC's view, reasonably ensure that funds will be available for decommissioning when needed. NRC-allowed funding assurance methods consist of two basic types -- licensee-funded and guarantees by a third party or, in certain circumstances, the licensee itself. Licensee-funded methods require a licensee to deposit funds or create reserves in either an external prepaid account or an external sinking fund accumulated over the reactor life in which the total amount of funds would be sufficient to pay decommissioning costs at the time that the plant is expected to permanently cease operations. The NRC requires that external funds be held by a trustee or escrow agent outside the licensee's control and segregated from its other assets. The NRC allows power reactor licensees that continue to be rate-regulated to collect decommissioning funds and deposit them into external trust accounts over the expected reactor life without providing any additional assurance of uncollected funds. Power reactor licensees that are no longer rate-regulated must provide additional assurance in the form of guarantees, as described below, for any unfunded portion of estimated decommissioning cost. As the electric utility industry is deregulated in the U.S., and as nuclear plants are sold and operated to run in a competitive environment, fewer licensees are rate-regulated. As a result, more licensees are required to ensure full funding of decommissioning. To date, all power plant sales to non-electric utilities have been accompanied by the prepayment of any unfunded decommissioning costs to meet the NRC's regulations.

The NRC believes that traditional cost-of-service ratemaking as it has been applied to "electric utilities" in the United States for most of the 20th century, coupled with the heavily capitalized nature of most nuclear electric generators, has provided a virtually assured source of funds for nuclear power plant construction, operation, and decommissioning. Nevertheless, the NRC recognizes that allowing power reactors that are "electric utilities" to accumulate funds in external sinking funds over the expected reactor life raises a potential problem of having inadequate funds in the event of a premature shutdown. The NRC evaluates such occurrences on a case-by-case basis. In all instances in which premature shutdown has entailed a funding shortfall, either the licensee's rate regulator has allowed the remaining funds to be collected from rate payers during the storage period before final dismantlement, or the licensee has had sufficient resources to make up any funding shortfalls. Similarly, although bankruptcy of licensees that are "electric utilities" is less likely because of rate regulation, bankruptcies that do occur might jeopardize future funds collection. However, the NRC has found that in the five instances in which power reactor licensees have filed for bankruptcy, they have filed as part of a reorganization under Chapter 11 of the U.S. Bankruptcy Code, rather than as a liquidation under Chapter 7. In the first four of these cases, the Bankruptcy Court has allowed the licensee to continue contributing to its operating and decommissioning costs for its share of the nuclear plant. In three cases, the bankrupt licensee emerged from bankruptcy without adversely affecting the status of decommissioning funding assurance for its plant. At the time of issuance of this report, the licensee for Diablo Canyon has filed for Chapter 11 protection, and the outcome has not yet been resolved.

The second basic funding assurance method is a guarantee in which a surety bonding company, insurer, or bank issuing a letter of credit promises to pay the licensee's decommissioning obligation if the licensee defaults. Additionally, pursuant to 10 CFR 50.75 (e)(2)(iii), if a licensee, or its parent company, is able to pass a financial test (see 10 CFR Part 30, Appendices A, C, and D), it may issue its own guarantee. Third-party guarantees may be used by themselves or in combination with external sinking funds. The NRC has had very little experience with third-party guarantees for power reactors. However, because of the limited market that currently exists for these mechanisms, and because of the face value amounts and time needed for them, it is unlikely that they would be available to most power reactor licensees. As the electric utility industry deregulates and fewer power reactor licensees meet the NRC's definition of "electric utility," and are thus eligible to use external sinking funds by themselves over the remaining reactor life, it is possible that third-party guarantee mechanisms will become more available.

Guidance regarding the form of trust agreements, guarantees, investments, and other issues germane to financial assurance mechanisms is provided in "Assuring the Availability of Funds for Decommissioning Nuclear Reactors," Regulatory Guide 1.159.

The NRC has closely monitored State and Federal deregulation initiatives because of their potential adverse effect on decommissioning funding assurance, and has initiated several actions to minimize any effects. Specifically, the NRC issued a "Final Policy Statement on the Restructuring and Economic Deregulation of the Electric Utility Industry" on August 19, 1997. This final policy statement addressed the NRC's public health and safety concerns about economic deregulation, including concerns about decommissioning funding assurance. The NRC also issued a final rule, entitled "Financial Assurance Requirements for Decommissioning Nuclear Power Reactors," which changed the agency's regulations to address deregulation-related changes in a licensee's ability to ensure decommissioning funding. Although it is too early to tell how rate deregulation will develop in all of the states that have nuclear power plants, those that have deregulated rates so far have recognized the importance of continued decommissioning funding assurance and have developed mechanisms (such as non-bypassable wires charges) that reasonably ensure that decommissioning funds will be available when, and in the amount, needed. The NRC has recognized the efficacy of such mechanisms in its regulations.

Decommissioning Funding Assurance for Nuclear Power Plants that have Permanently Shut Down

Once a licensee has permanently shut down its nuclear power plant, its decommissioning activities, including those that are funding related, are governed by 10 CFR 50.82, "Termination of License." As indicated above, 10 CFR 50.82(c) provides that, for a facility that has permanently ceased operation before the expiration of its license, the collection period of any shortfall of funds will be determined upon application by the licensee, on a case-by-case basis, taking into account the specific financial situation of each licensee.

No later than 2 years following permanent cessation of operations, each power reactor licensee is required to submit to the NRC a post-shutdown decommissioning activities report. Among its contents, this report must describe the planned decommissioning activities, a schedule for their accomplishment, and an estimate of expected costs. Licensees are prohibited from performing any major decommissioning activities until 90 days after the NRC has received the licensee's report, and until the licensee submits certifications of both permanent cessation of operations and permanent removal of fuel from the reactor vessel. Licensees are also prohibited from performing any decommissioning activities that could challenge the reasonable assurance that adequate funds will be available for decommissioning.

The NRC has also stipulated standards for withdrawing funds from external trust accounts. Specifically, 10 CFR 50.82(a)(8) provides that a licensee may use such funds if the following criteria are met:

(A) The withdrawals are for expenses for legitimate decommissioning activities consistent with the definition of decommissioning in 10 CFR 50.2, "Definitions."

(B) The expenditure would not reduce the value of the decommissioning trust below an amount that is necessary to place and maintain the reactor in a safe storage condition if unforeseen conditions or expenses arise.

(C) The withdrawals would not inhibit the ability of the licensee to complete funding of any shortfalls in the decommissioning trust needed to ensure the availability of funds to ultimately release the site and terminate the license.

10 CFR 50.82 also contains criteria for timing trust fund withdrawals if a licensee has met the preceding standards. A licensee may withdraw up to 3 percent of the generic formula amount specified in 10 CFR 50.75(c) at any time, even before reactor shutdown, for decommissioning planning. Once a licensee has submitted the certifications of permanent cessation of operations and permanent removal of the fuel from the reactor vessel, and commencing 90 days after the NRC has received the post-shutdown decommissioning activities report, the licensee may withdraw an additional 20 percent. A licensee may use the remaining funds in its external trust account once it submits a detailed, site-specific cost estimate to the NRC.

Finally, each licensee who has formulated a decommissioning plan that delays completion of decommissioning by including a period of storage or surveillance must ensure that the funds needed to complete decommissioning will either (1) be placed into an account segregated from the licensee's assets and outside the licensee's administrative control during storage or surveillance, or (2) maintain a surety method in accordance with the criteria of 10 CFR 50.75(e). The licensee must also demonstrate a means to adjust cost estimates and associated funding levels over the storage or surveillance period.

11.1.2.2 Experience and Examples

Several U.S. power reactors have permanently shut down. Some have completed decommissioning and terminated their NRC licenses; others are in various stages of decommissioning. The licensees of many shutdown plants have decided to defer final dismantlement. This is generally the case at multi-unit sites where other reactors continue to operate. Licensees that have decommissioned have incurred costs for NRC-required decommissioning activities ranging from $180 million to $500 million. However, these plants have been smaller, uniquely designed, or with short operational lives; so their decommissioning costs may not be representative of large light-water reactors that have operated for several decades. The Trojan Plant, an 1130-MW(e) pressurized-water reactor is currently undergoing dismantlement. The NRC expects that the experience gained over the next few years from the dismantlement activities at Trojan and other light-water reactors will provide additional data regarding the range of decommissioning costs that are likely to be incurred at a large light-water reactor.

11.1.3 Financial Protection Program for Liability Claims Arising From Accidents

This section explains the financial protection program for liability claims arising from accidents. It covers the governing documents, primarily the Price-Anderson Act, and process to implement requirements. It also discusses relevant experience and gives examples.

11.1.3.1 Governing Documents and Process

The Price-Anderson Act, enacted in 1957, became Section 170 of the Atomic Energy Act of 1954, as amended, and governs the U.S. financial protection program. Section 170 (with related definitions in Section 11) provides the financial and the legal framework to compensate those who suffer bodily injury or property damage as a result of accidents at covered nuclear facilities. The NRC's regulations implementing the provisions of Section 170 for NRC licensees are codified in 10 CFR Part 140, "Financial Protection Requirements and Indemnity Agreements."

The Price-Anderson Act was enacted to meet two basic objectives:

(1) Remove the deterrent to private sector participation in atomic energy presented by the threat of potentially enormous liability claims in the event of a catastrophic nuclear accident.

(2) Ensure that adequate funds are available to the public to satisfy liability claims if such an accident were to occur.

In enacting the Price-Anderson Act, the U.S. Congress sought an equitable balance of industry's needs with those of the public. Specifically, Congress required that all reactor licensees (sometimes "operators"), among other licensees, purchase specified amounts of liability insurance (then at a maximum level of $60 million) or possess other equal financial protection against the risk of a nuclear accident. Beyond the financial protection, the U.S. government agreed to indemnify the licensees for each accident for an additional $500 million of damages in order to achieve reasonable compensation for the public; it also agreed to limit the total liability for an accident. The limit was set at the sum of the required financial protection and the government indemnity (thus a maximum limit of $560 million which was applicable largely to commercial power reactors.)

The financial protection, indemnification and liability limit applied not only to the liability of the licensee but also to the aggregate sum of all liability for all persons who might be held liable. This "omnibus coverage" effectively channeled the financial responsibility for all damages up to the liability limit to the licensee and, in turn to the government. In so providing, Congress indemnified the suppliers, contractors and others in the nuclear power industry, as needed in the event of an accident, and assured the availability of reasonable compensation no matter what caused the accident.

The Price-Anderson Act has been revised several times since 1957, the most recent being in 1988 when Congress renewed the Commission's authority to cover new facilities until August 2002. As revised over the years, the means for providing financial protection for power reactors over 100 MWe have changed significantly. Under current law, those reactors must contribute to a pool that replaces the government as the second provider of funds if the first layer of financial protection (liability insurance--now $200 million) is exhausted.

The reactor operators are required after an accident to pay into a "retrospective premium pool," in maximum annual installments not exceeding $10 million, up to a total of $83.9 million each. But payment is called for only if the accident exhausts the first layer of financial protection, and only if and to the extent that additional funds are needed to pay the damages. With 106 reactors currently participating in the system (104 are power reactors with operating licenses), the total financial protection available under the Price-Anderson Act for any one accident is $9.09 billion [$200 million primary coverage + (83.9 million per reactor x 106 reactors)]; $9.09 billion is also the limit on liability. As reactors leave the retrospective premium system as a result of permanent closure or join as the result of construction of new reactors, this coverage limit may fall or rise. A change in the limit may also occur when the $83.9 million contribution is adjusted for inflation, as must be done every five years. In any event, Congress will address any damages exceeding the total sum that reactors must contribute to the pool and will decide upon the next steps needed for compensation.

The public is significantly benefitted by another feature of the Act. Claimants need only prove that the accident caused their injury in order to receive compensation for damages from any accident with significant offsite releases of radiation, i.e. an "extraordinary nuclear occurrence." No proof of fault is necessary, or even of what caused the accident.

Section 170 itself provides greater detail such as designation of the federal district court to try claims cases, application of limits of liability, coverage of other licensees, permissible claims (including injuries caused by sabotage, coverage for precautionary evacuations, settlement expenses, and punitive damages outside government indemnification). Such detail exceeds the scope of this discussion. A more detailed summary is available in the NRC publication, "The

Price-Anderson Act–Crossing the Bridge to the Next Century: Report to Congress,"
NUREG/CR-6617.

11.1.3.2 Experience and Examples

Claims for more than 100 alleged incidents involving nuclear material have been filed under various liability policies since the inception of the Price-Anderson Act in 1957. Earlier claims tended to be property damage claims arising from alleged radiation from leakage or from other accidents involving containers of nuclear materials in transit. More recent claims have emphasized bodily injury arising from alleged radiation exposure, especially by contractor employees working at the sites of operating nuclear power plants. The insured losses and expenses paid so far total over $80 million. Of this amount, most payments arose out of the accident at Three Mile Island Unit 2 (TMI-2). The insurers responded rapidly to the accident, and established an office to pay claims for the living expenses of the families with pregnant women and pre-school-age children who evacuated the 5-mile area around TMI- 2. Following the accident, numerous lawsuits were filed, alleging various injuries and property damages. The lawsuits were consolidated into one suit and, in September 1981, a settlement agreement was signed. Under the terms of the agreement, the insurers paid into a Court-managed fund $20 M for economic harm to businesses and individuals within 25 miles of the plant, and $5 M to establish a Public Health Fund in the TMI- 2 area.

11.1.4 Insurance Program for Onsite Property Damages Arising From Accidents

This section explains the insurance program for onsite property damages arising from accidents. It covers the governing documents, the process to implement requirements, relevant experience and examples.

11.1.4.1 Governing Documents and Process

Among other sections of the AEA, Section 182.a provides the basis for the NRC's onsite property damage insurance requirements for operating nuclear power reactors contained in Paragraph (w) of 10 CFR 50.54, "Conditions of Licenses."

After the accident at TMI-2, the NRC evaluated the effect of accidents on the costs of decommissioning. The resulting study, "Technology, Safety, and Costs of Decommissioning Reference Light-Water Reactors Following Postulated Accidents," NUREG/CR-2601, published in 1982, evaluated three accident scenarios and the probable effect of each on decommissioning costs. In addition to its effect on decommissioning costs, experience at TMI-2 indicated that the potential financial impact on the licensee/owner of the plant suffering the accident could be substantial, to the point that a licensee might not have adequate funds to stabilize or decontaminate the reactor. (At the time of the accident, that plant's owners carried onsite insurance of $300 M, the maximum then available. Total onsite cleanup costs at TMI-2 exceeded $1 B, which caused the plant's owners serious financial stress.) Thus, the NRC became concerned that inadequate insurance or equivalent financial resources that could be devoted to the reactor site itself after an accident could have a negative effect on the protection of public health and safety.

In view of this concern, the NRC adopted regulations requiring its power reactor licensees to obtain onsite property damage insurance, or an equivalent source of protection, that would be used, before any other purpose, to ensure that the reactor is maintained in, or is returned to, a safe and stable condition, and that radioactive contamination is removed or controlled so that personnel exposures are consistent with the occupational exposure limits in 10 CFR Part 20, "Standards for Protection Against Radiation." Actions needed to return the reactor to and maintain it in a safe and stable condition could include one or more of the following steps, as appropriate:

- Shutdown the reactor.

- Establish and maintain long-term cooling with stable decay heat removal.

- Maintain subcriticality.

- Control radioactive releases.

- Secure structures, systems, or components to minimize radiation exposure to onsite personnel or the offsite public, or to facilitate later decommissioning, or both.

The NRC specified that this coverage had to be at least $1.06 B (the amount derived from the agency's study). Available insurance exceeded this amount by the time these NRC requirements went into effect. Most licensees purchase the maximum onsite insurance currently available (about $3 B). In the event of an accident, licensees would be able to use the proceeds from this insurance above the NRC-required amount, for example, to pay for the costs to replace equipment that is damaged in the accident.

11.1.4.2 Experience and Examples

The U.S. nuclear industry has not experienced an accident of significant radioactive contamination since TMI-2. Thus, the NRC has not needed to invoke the stabilization and cleanup provisions of its onsite insurance programs since these requirements went into effect.

11.2 Regulatory Requirements for Qualifying, Training, and Retraining Personnel

This section explains the regulatory requirements for qualifying, training, and retraining personnel. It discusses the governing documents, process for implementing requirements, experience, and examples.

11.2.1 Governing Documents and Process

The general regulation in 10 CFR 50.40, "Common Standards," ensures that licensees must demonstrate that they are technically qualified to engage in nuclear activities. Paragraph (b)(6)(i) of 10 CFR 50.34, "Contents of Applications," requires information about personnel qualifications, while, 10 CFR 50.54, "Conditions of Licenses," contains specific requirements regarding facility operators.

Issued in 1987, 10 CFR Part 55, "Operator Licensing," regulates the training requirements for licensed operators and licensed senior operators, while allowing facility licensees to have operator requalification program content that is derived using a systems approach, or that meets the requirements outlined in paragraph (c)(1) of 10 CFR 55.59, "Requalification." Subpart D, "Applications," of 10 CFR Part 55 requires that operator license applications must contain information about an individual's education and experience.

The operator licensing process at power reactors (as discussed in 10 CFR Part 55) includes a generic fundamentals examination, which covers the theoretical knowledge that is required to operate a nuclear power plant, as well as a site-specific examination, which consists of a written examination and an operating test that includes a plant walkthrough and a dynamic performance demonstration on a simulation facility. License applicants must pass the generic fundamentals examination before they can take the site-specific examination.

The NRC staff has worked to redefine the operator licensing function to transfer additional responsibility for developing examinations to licensees. The NRC issued a final rule in April 1999, amending 10 CFR Part 55, "Operators' Licenses," to allow nuclear power reactor licensees to prepare the written examinations and operating tests that the NRC uses to evaluate the competence of applicants for operators' licenses at those facilities. Licensees that elect to prepare their own examinations are required to establish and implement procedures to control examination security and integrity and to prepare and submit proposed examinations and operating tests to the NRC according to the guidance in NUREG-1021, "Operator Licensing Examiner Standards," Revision 8, Supplement 1 issued in April 2001. The NRC will review facility-prepared examinations, and will continue to administer all operating tests and make the final licensing decisions.

In response to a 1990 court decision, in 1993, the NRC issued 10 CFR 50.120, "Training and Qualification of Nuclear Power Plant Workers." This rule requires that training programs be established, implemented, and maintained using a systems approach to training for eight categories of nonlicensed workers at nuclear power plants and the shift supervisor who is licensed. 10 CFR 50.120 complements the requirements for training based on a systems approach for the requalification of licensed operators.

As guidance to implement the requirements of the regulations, the NRC issued Regulatory Guide 1.8, "Personnel Selection and Training." This guide states the NRC's position on plant staff personnel qualifications (both licensed and nonlicensed), and indicates that the criteria for the qualifications contained in ANSI/ANS-3.1, "Selection and Training of Nuclear Power Plant Personnel," are generally acceptable.

The NRC monitors industry performance in implementing the training requirements of 10 CFR Parts 50 and 55 by (1) reviewing licensee event reports and inspection reports for training issues, (2) observing the accreditation process, and (3) reviewing the results of operator licensing activities. Guidance for inspecting the aspects of the operator training programs that are unique to requalifications is given by Inspection Procedure 71111.11, "Licensed Operator Requalification Program." In addition, the NRC verifies compliance with the requirements for training based on a systems approach through its inspection program when appropriate for cause, using Inspection Procedure 41500, "Training and Qualification Effectiveness," which references the guidance in NUREG-1220, Revision 1 "Training Review Criteria and Procedures," issued in January 1993.

In accordance with its memorandum of agreement with the Institute of Nuclear Power Operations (INPO), the NRC monitors INPO accreditation activities as part of its assessment of the effectiveness of the industry's training programs. (The NRC also monitors the selected performance areas of its licensees as part of its assessment.) The NRC monitors INPO activities by observing accreditation team visits and the monthly National Nuclear Accrediting Board meetings. These visits are intended to monitor the implementation of programmatic aspects of the accreditation process.

Placing a training program on probation or withdrawing accreditation indicates a Board concern. It does not necessarily place a training program in noncompliance with either 10 CFR 50.55 or 10 CFR.120, since training programs are accredited to a "standard of excellence" rather than a minimum level of regulatory compliance. However, the NRC does review the circumstances leading to the withdrawal or probation to ensure safe operations and continued compliance with regulations.

The Board may withdraw accreditation in response to major deficiencies in a licensee's accredited training program. If accreditation is withdrawn, the NRC would request that the licensee report the circumstances of the withdrawal for the staff to determine the significance of the issues related to the withdrawal. If the NRC determines that compliance with the regulations is not affected, the agency may not need to take any further action. If the withdrawal is linked to a breakdown in the training process or a safety-significant issue, the agency will conduct an immediate inspection focused on the process problem or safety issue(s). It would take further action, such as issuing Orders, if appropriate.

11.2.2 Experience and Examples

The NRC reviewed training issues contained in licensee event reports and inspection reports at the end of 1999 using data from the Human Factors Information System. (This system is also described in Article 12.) This review revealed that, over the past 3 years, the contribution of training has remained relatively constant at about seven percent for the industry as a whole. The identified training issues continue to be concentrated in two distinct areas-- training less than adequate" and "individual knowledge less than adequate." The declining trend in "training less than adequate" is in contrast to the increasing trend in "individual knowledge less than adequate." In other words, the data suggest that the causes of poor performance appear to be becoming more focused on the individual, rather than on a group or class of worker. The worker knowledge deficiencies may be a reflection of the continuing systems approach to training process problems, which is also indicated by the data. Given the maturity level of most industry training programs, the process elements on the evaluation of trainee evaluation and program effectiveness are the elements that are most closely linked to performance deficiencies.

In monitoring industry performance in training, the NRC concluded that INPO accreditation continues to be an acceptable means of ensuring that the training requirements in 10 CFR Parts 50 and 55 are being met. Although NRC monitoring gives some indications of limited specific weaknesses in training programs, all indicators suggest that the industry is successfully implementing training programs in accordance with the regulations. Monitoring of selected performance areas will continue with emphasis placed on identifying training process problems and ensure that they are appropriately resolved.

An example of this monitoring process is a "for cause" inspection of the training of licensed operator candidates at the St. Lucie plant in March 1999. That inspection assessed the licensee's response to the low pass rate on the initial licensing examination given in December 1998. As part of that inspection, the inspectors interviewed a number of licensee personnel, including line supervisors, training management personnel, instructors, and licensed operators. They also observed classroom training sessions and reviewed meeting minutes, training procedures, and course evaluations. The inspectors found that an analysis of changes to the schedule used to implement the initial licensed operator training program had not been adequate, and that the licensee did not update some of the lesson material and did not sufficiently consider trainee critiques when evaluating programs.

ARTICLE 12. HUMAN FACTORS

Each Contracting Party shall take the appropriate step to ensure that the capabilities and limitations of human performance are taken into account throughout the life of a nuclear installation.

This section explains the NRC's program on human performance and the 6 major areas under the program in which the NRC performs significant activities. These areas are (1) human factors engineering issues; (2) emergency operating procedures and plant procedures; (3) working hours and staffing; (4) fitness-for-duty; (5) human factors information system; (6) support to event investigations and for-cause inspections; and (7) training. This section also discusses research activities.

12.1 Background

Human performance is a critical element of nuclear power plant safety. More than half of the incidents that are reported by licensees of commercial nuclear power plants have human performance as a root cause. The importance of the role of human performance in the safe use of nuclear materials is underscored by the finding of the International Nuclear Safety Advisory Group (as reported in INSAG-3, "Basic Safety Principles for Nuclear Power Plants," 1988) that many incidents at nuclear power plants often result from incorrect human actions. Consistent with the NRC's mission is the view that human performance should not contribute to undue risk in using nuclear materials.

12.2 NRC Program on Human Performance

This section discusses the NRC Program on Human Performance and significant activities performed under this program.

12.2.1 Goals and Mission of the Program

The NRC has a comprehensive program for ensuring that human performance is properly addressed in a risk-informed regulatory framework for maintaining reactor safety. The agency developed the program based on reviewing risk information, as well as information from sources such as activities in the domestic and international nuclear industry. SECY-00-0053, "NRC Program on Human Performance in Nuclear Power Plant Safety," dated February 29, 2000, describes the program in detail. The mission of the program is to ensure that reactor safety is maintained through effective regulation and oversight of human performance in the design, operation, maintenance, and decommissioning of nuclear reactor facilities. The NRC accomplishes this mission through this program by: (1) identifying human performance issues important to public health and safety, (2) increasing its understanding of the causes and safety implications of human performance issues, and (3) properly responding. The goals of this program are (1) to ensure that NRC human performance activities are directed at improving the understanding of risk significance, (2) to improve the modeling of human performance, and (3) to use risk insights to inform regulatory activities.

12.2.2 Program Elements

The program applies to the Revised Reactor Oversight Process, plant licensing and monitoring, the Risk-Informed Regulation Implementation Plan, and emerging technology and emerging issues.

The Revised Reactor Oversight Process (discussed in Article 6) focuses on cornerstones of safety which are assessed through a combination of performance indicators and risk-informed inspections. The inspections focus on risk-significant activities and systems related to the cornerstones. There are three elements that are considered crosscutting to the cornerstones: human performance, safety-conscious work environment, and corrective actions. The Program on Human Performance has contributed directly to the development of a supplemental inspection procedure related to the human performance crosscutting element. This program also contributes indirectly to the other two elements, since a safety conscious work environment is a human performance element and many of the actions involved in corrective action programs result from human performance problems.

Activities that apply to plant licensing and monitoring include reviewing licensing actions and monitoring plant and program performance. As part of regulatory initiatives, the staff reviews Commission policies related to human performance that address identified problems. The staff also supports rulemaking and the development of regulatory guidance.

Support to the Risk-Informed Regulation Implementation Plan includes generating, collecting, and evaluating data on human performance for use in human reliability analysis models. The NRC staff evaluates information to gain insights to support risk-informed regulation and to provide human performance data for human reliability analysis.

Activities that apply to emerging technology and emerging issues are intended to prepare the NRC for the future. The two activities applying to these categories are developing regulatory guidance for reviewing designs of control stations and processing requests related to deregulation. Licensees are replacing aging analog controls and displays with digital components, and the agency needs to be prepared to review safety issues that arise from human-system interfaces resulting from such new designs and technologies. With regard to deregulation, the NRC has been processing numerous industry requests to transfer operating licenses, which may involve changes in organizational structure affecting human performance as discussed further below.

12.2.3 Significant Regulatory Activities

The NRC performs significant regulatory activities in the following areas to address human performance under the program:

(1) Human Factors Engineering Issues

(2) Emergency Operating Procedures and Plant Procedures

(3) Working Hours and Staffing

(4) Fitness-for-Duty

(5) Human Factors Information System

(6) Support to Event Investigations and For-Cause Inspections

(7) Training

The first six are described below. Training is described under Article 11, "Financial and Human Resources."

12.2.3.1 Human Factors Engineering Issues

This section discusses human factors activities related to engineering issues. It covers the governing documents and process to carry out requirements, and experience and examples.

Governing Documents and Process

The NRC staff evaluates the human factors engineering design of the main control room and control centers outside of the main control room using NUREG-0800, Revision 1, Chapter 18, "Human Factors Engineering." It also uses Revision 1 to NUREG-0700, "Human System Interface Design Review Guideline," issued in 1998. NUREG-0700 provides guidance on human factors engineering to the NRC staff for its reviews of submittals on human-system interface design related to licenses or design certification of nuclear installations, and reviews of the human-system interface that could be performed as part of an inspection.

Experience and Examples

The NRC reviews licensees' requests that involve aspects of human factors engineering. Examples include crediting manual actions in amendments to plant technical specifications, license transfers, and increasing the reactor power level ("power uprates").

As an example, the staff recently evaluated requests for license amendments from Byron/Braidwood stations and the Callaway plant. These licensees requested credit for manual action instead of automatically actuated systems to mitigate an inadvertent boron dilution event. The NRC used information in Information Notice 97-78, "Crediting of Operator Actions in Place of Automatic Actions and Modifications of Operator Actions, Including Response Times," as its principal review source for such requests.

The NRC has evaluated a number of requests to transfer operating licenses, paying special attention to management and organization, staffing, and technical qualifications. The staff uses NUREG-0800, Standard Review Plan, Chapter 13, "Conduct of Operations," as principal guidance for these reviews. The staff revised this chapter in 1999 to better address license transfer reviews.

The NRC has also approved requests for power uprates of currently licensed plants. For such requests, the staff examines the effect of the proposed power uprate on plant procedures, controls, displays and alarms, and required operator actions.

12.2.3.2 Emergency Operating and Plant Procedures

This section discusses human factor activities related to emergency operating and plant procedures. Licensees must have programs for developing, implementing and maintaining such procedures.

Governing Documents and Process

Generic Letter 82-33, "Requirements for Emergency Response Capability" (which transmitted Supplement 1 to NUREG-0737, "Requirements for Emergency Response Capability"), requires each licensee to submit a set of documents for developing emergency operating procedures.

The NRC staff conducts for-cause inspections of plant procedures and emergency operating procedures using Inspection Procedures 42700, "Plant Procedures," and 42001, "Emergency Operating Procedures," respectively. The staff uses Inspection Procedure 42700 to focus inspections on identified procedural problems. In particular, this procedure guides inspectors on inspecting the usability of a licensee's procedures by assessing the degree to which accepted human factors principles have been incorporated into them. Inspection Procedure 42001 gives guidance for inspecting emergency operating procedures -- their development, implementation, revisions, and maintenance. Inspections include onsite human factors specialists and systems experts.

Experience and Examples

No significant examples applying to emergency operating and plant procedures were identified since 1998.

12.2.3.3 Working Hours and Shift Staffing

This section discusses activities related to working hours and shift staffing. It covers the governing documents and process to implement requirements and gives experience and examples.

Governing Documents and Process

Working Hours

The NRC's policy on working hours is stated in the "Policy on Factors Causing Fatigue of Operating Personnel at Nuclear Reactors," dated February 18, 1982, and revised in June 1982. The objective of the policy is to ensure (to the extent practicable) that personnel are not assigned to shift duties while in a fatigued condition that can significantly reduce their mental alertness or decisionmaking ability. The policy applies to "unit staff who perform safety-related functions (e.g., senior reactor operators, reactor operators, auxiliary operators, health physicists, and key maintenance personnel)," and it gives specific guidance to be used if unforeseen problems require substantial amounts of overtime. The policy also allows deviations from the guidelines "for very unusual circumstances" provided that (1) the plant manager, his deputy, or higher levels of management authorize the deviations, and (2) it would be "highly unlikely" that such deviations would cause significant reductions in the effectiveness of operating personnel.

In March 1983, the NRC issued Generic Letter 83-14, "Definition of Key Maintenance Personnel (Clarification of Generic Letter 82-12)," to clarify the applicability of the policy to maintenance personnel. Specifically, Generic Letter 83-14 defined key maintenance personnel as "those personnel who are responsible for the correct performance of maintenance, repair, modification, or calibration of safety-related structures, systems, or components, and who are performing or immediately supervising the performance of such activities."

Shift Staffing

Paragraph (m) of 10 CFR 50.54, "Conditions of Licenses," specifies the minimum number of licensed operators that are required for nuclear power reactor sites. The required number of reactor operators and senior reactor operators depends, in part, upon the number of units, the operating mode of the units, and the number of control rooms (i.e., whether multi-unit sites use common or independent control rooms). The most recent amendment to 10 CFR 50.54(m) increased the minimum onsite staffing for a single operating unit to two senior reactor operators and two reactor operators, a senior reactor operator being in the control room at all times and a reactor operator or additional senior reactor operator at the controls when the plant is in an operating mode. In addition to the shift staffing requirements of 10 CFR 50.54(m), the NRC has other requirements with staffing implications. These include the personnel requirements for fire brigades and emergency response personnel contained in Appendix R, "Fire Protection Programs for Nuclear Power Facilities Operating Prior to January 1, 1979," and Appendix E, "Emergency Planning and Preparedness for Protection and Utilization Facilities," to 10 CFR Part 50, respectively.

Experience and Examples

Working Hours

Most of the NRC's inspection findings regarding work scheduling only addressed administrative weaknesses in the control of work hours. However, the NRC has occasionally identified more significant work scheduling concerns. In February 1999, Congressional correspondence expressed concern that low staffing levels and excessive overtime could present a serious safety hazard at some commercial nuclear power reactors. Similar concerns were expressed in a Union of Concerned Scientists report, issued in March 1999, which addressed "Overtime and Staffing Problems in the Commercial Nuclear Power Industry." The NRC staff preliminarily reviewed inspection reports and licensee event reports from 1994 through April 1999. The review revealed that few events at nuclear power reactors had been attributed to fatigue and, in all cases, automated safety systems or other barriers were available to prevent events that could have had safety consequences. However, the staff acknowledged that the number of events that were attributable to fatigue could not be reported with certainty, given the difficulty of making such determinations, and that NRC inspectors had identified several instances each year in which licensees' use of overtime appeared to be inconsistent with the general objectives or specific guidelines of the NRC's policy statement. As a result, the NRC staff committed to assess the policy more comprehensively.

In September 1999, the Commission received a petition for rulemaking (PRM-26-2), which requested that the NRC establish clear and enforceable work hour limits to mitigate the effects of fatigue for nuclear power plant personnel performing safety-related work. The NRC subsequently formed a working group to concurrently assess the policy and respond to the petition. In addition, the Nuclear Energy Institute (NEI) surveyed guideline deviations at U.S. nuclear power plants. The survey indicated that most of the sites authorized few deviations from the NRC's guidelines. However, the survey also indicated that (1) about one-third of the survey respondents authorized 1000 - 7500 approvals in a year that exceed the NRC's policy guidelines, (2) 8 of 36 sites providing data had more than 20 percent of the personnel covered by the policy working in excess of 600 hours of overtime per year, and (3) a limited number of sites may not be applying work hour controls to all personnel who perform safety-related functions. The NRC is currently reviewing the working group's assessment of the policy, its implementation, and the petition for rulemaking to determine if changes are warranted in the NRC's regulatory requirements and guidance regarding fatigue and the control of work hours.

Shift Staffing

No significant examples of shift staffing were identified from 1998 through 2000.

12.2.3.4 Fitness-for-Duty

The NRC first published a rule titled, "Fitness-for-Duty Programs," in 1989. The rule required each licensee authorized to operate or construct a nuclear power reactor to implement a Fitness-for-Duty Program for all personnel having unescorted access to the protected area of its plant. The rule specified, as a performance objective, that licensees must provide reasonable assurance that nuclear power plant personnel perform their tasks in a reliable and trustworthy manner and are not under the influence of any substance, legal or illegal, or mentally or physically impaired from any cause. In addition, the rule required that licensee policy should also address factors that could affect fitness for duty such as mental stress, fatigue, and illness. On December 4, 2000, the Commission approved a revision to the rule that is presented in SECY-00-0159, "Final Rule Amending the Fitness-for-Duty Rule." Currently, the rule is before the Commission along with the NRC's staff positions and recommendations with respect to stakeholder's concerns regarding implementing the final rule.

12.2.3.5 Human Factors Information System

This section describes the Human Factors Information System and discusses experience and examples.

Description of the System

The Human Factors Information System is designed to be a convenient means of storing, retrieving, sorting, and analyzing human performance information. An automated management information system, it contains human performance information extracted from inspection reports, operator licensing examination reports, and licensee event reports. The system also contains information regarding the status of the accredited training programs at each site. The system can generate a variety of specialized reports, and has built-in system maintenance functions. Initiated in 1990, the system compiles human factors information that is not readily available from other NRC sources. Sources include NRC inspections and audits at plant sites. The agency maintains a web page, (http://www.nrc.gov) to disseminate information on human performance issues at individual nuclear power plant sites.

The staff uses information from the system to gain insights about human performance and to monitor the frequency of human performance occurrences related to staffing, training, overtime, procedures, and human-system interface. The system is used in preparing for plant performance assessments. In such instances, human performance data are provided for the most recent 12-month period for each region. The total number of human performance factors contributing to each plant's licensee event report during the reporting period is compared to the national average. Written analyses of the data, including the types of personnel and the performance issues, are prepared for plants that meet or exceed twice the national average (a threshold value). In addition, the data for previous years are presented in tabular form for trending. Written analyses are also provided when plants do not meet the threshold but have shown a significant increase or decrease in the number of human performance-related contributing factors in licensee event reports.

Experience and Examples

The written analyses provided in preparing for plant performance assessments (described above) include the following recent examples:

- An analysis of the Procedures and Reference Document category for a plant indicated that the percent contribution in this area is less than the industry profile. However, 38 of the 43 procedure items (88 percent) were the result of human factors deficiencies or incorrect procedure technical content. About one-third of the procedure items were attributed to the engineering department. The items were identified throughout the course of the year.

- An analysis of the Work Factors category for a plant indicated that the percent contribution in this area is less than the industry profile. However, two related issues were major contributors. Specifically, "less than adequate work package quality" and "less than adequate work practice or skill of the craft" accounted for about one-third of the 148 items. These issues were identified throughout the course of the year and were attributable to several departments. However, the engineering department was the largest contributor.

- Direct supervisory oversight was involved in 57 of the 144 (40 percent) management and supervisory items at a plant. Inadequate supervisory command and control was involved in 67 items (47 percent).

- A plant was 2 percent higher than the industry average in the category of Procedures and Reference Documents. An analysis of this category indicated that 32 of the 35 procedure and reference document items (92 percent) identified problems with the content of procedures. These items were evenly split between the technical content issues and human factors deficiencies. Inadequate technical content in procedures was identified as being a contributor to five separate licensee event reports.

- A plant was 84 percent higher than the national average in the category of Management and Supervision. An analysis of this category indicated that 13 of the 33 management and supervision items were in the corrective action area of "action not begun or untimely," and 8 of the 33 management and supervision items were in the corrective action area of "individual corrective action less than adequate." More than three-quarters of these items were concentrated in electrical maintenance and in the engineering department.

It is important to note that the information provided was somewhat subjective, and was only used to give insight or validate information from other sources.

12.2.3.6 Support to Event Investigations and For-Cause Inspections and Training

Human factors technical expertise is included in special inspections; incident investigation team inspections; augmented inspection team inspections; event investigations; and projects, programs, and policy activities. The NRC staff assesses management effectiveness, procedures, training issues, staffing issues, and human-machine interfaces. For these activities, inspectors use Inspection Procedure 41500, "Training and Qualification Effectiveness" to ensure that training and qualification programs for nuclear power plant personnel are developed, carried out, evaluated, documented, and maintained as required by 10 CFR 50.120 and allowed by 10 CFR Part 55. For baseline inspections under the Revised Reactor Oversight Process, inspectors use Inspection Procedure 71152, "Identification and Resolution of Problems." The procedure is intended to establish confidence that each licensee is detecting and correcting problems in a manner that limits the risk to the public. A key premise of the Revised Reactor Oversight Process is that weaknesses in licensees' problem identification and resolution programs will manifest themselves as performance issues that can be identified during the baseline inspection program or by crossing predetermined indicator thresholds.

12.3 Significant Research Activities

In addition to its regulatory activities, the NRC researches human performance issues. This research has resulted in the publication of a number of reports in 2000. NUREG/CR-6689, "A Proposed Approach for Reviewing Changes to Risk-Important Human Actions," provides a risk-informed methodology for use in evaluating changes to the licensees' design bases that affect human performance. Several NUREG/CRs address the effects of upgrades that are being made to plant systems and the human-system interface on the performance of plant personnel. The NRC is currently sponsoring research to address the potential effects that advances in technology (such as converting from analog to digital systems) might have on human performance. Ultimately, the goal is to develop guidance and review criteria to support safety reviews. These reports include the following:

- NUREG/CR-6637, "Human Systems Interface and Plant Modernization Process: Technical Basis and Human Factors Review Guidance."

- NUREG/CR-6635, "Soft Controls: Technical Basis and Human Factors Review Guidance."

- NUREG/CR-6634, "Computer-Based Procedure Systems: Technical Basis and Human Factors Guidance."

- NUREG/CR-6636, "Maintainability of Digital Systems: Technical Basis and Human Factors Guidance."

- NUREG/IA-0137, "A Study of Control Room Staffing Levels for Advanced Reactors."

- NUREG/CR-6691, "The Effects of Alarm Display, Processing, and Availability on Crew Performance."

- NUREG/CR-6684, "Advanced Alarm Systems: Revision of Guidance and Its Technical Basis."

- NUREG/CR-6633, "Advanced Information System Design: Technical Basis and Human Factors Review Guidance."

ARTICLE 13. QUALITY ASSURANCE

Each Contracting Party shall take the appropriate steps to ensure that quality assurance programmes are established and implemented with a view to providing confidence that specified requirements for all activities important to nuclear safety are satisfied throughout the life of a nuclear installation.

This section explains quality assurance (QA) policy and requirements, and guidance for design and construction, operational activities, and staff licensing reviews. It also describes QA programs, including QA under the Revised Reactor Oversight Process, augmented QA, and graded QA.

13.1 Background

Nuclear power facilities must be designed, constructed, and operated in a manner that ensures (1) the prevention of accidents that could cause undue risk to the health and safety of the public, and (2) the mitigation of adverse consequences of such accidents if they should occur. A primary means for achieving these objectives is by establishing and effectively implementing a nuclear QA program. Although a licensee may delegate aspects of the establishment or execution of the QA program to others, the licensee remains ultimately responsible for its overall effectiveness of the program. Licensees perform a variety of self-assessments to validate the effectiveness of their QA program implementation. The NRC reviews descriptions of QA programs, and performs onsite inspections to verify aspects of the program implementation.

13.2 Regulatory Policy and Requirements

Each applicant for a construction permit for a nuclear power plant is required by paragraph (a)(7) of 10 CFR 50.34 "Contents of Applications; Technical Information," to describe its QA program in its preliminary safety analysis report. This program applies to the design, fabrication, construction, and testing of safety-related plant equipment. Each applicant for a license to operate a nuclear power plant, is required by paragraph (b)(6)(ii) of 10 CFR 50.34 to provide a final safety analysis report that details its managerial and administrative controls to ensure safe operation. In both reports, the applicant must describe how it will satisfy the applicable requirements of 10 CFR Part 50, Appendix B, "Quality Assurance Criteria for Nuclear Power Plants and Fuel Reprocessing Plants."

If a licensee wants to make changes in its QA program, it is required by paragraph (a)(3) of 10 CFR 50.54 "Conditions of Licenses" to inform the NRC of the changes. A licensee can make changes without prior NRC approval if the changes do not reduce the commitments in the program description as accepted by the NRC. In April 1999, the NRC revised 10 CFR 50.54(a) to define six categories of QA Program changes that are not "reductions in commitments." Changes that do reduce commitments related to the QA Program must receive NRC approval before implementation.

Nuclear quality assurance criteria apply to all activities that affect the safety-related functions of structures, systems, and components that prevent or mitigate the consequences of postulated accidents that could cause undue risk to the health and safety of the public. High-level criteria for determining which plant structures, systems, and components are safety related are provided in 10 CFR 50.2, "Definitions". The definition is consistently applied in Appendix A to 10 CFR Part 100, "Seismic and Geologic Siting Criteria for Nuclear Power Plants," 10 CFR 50.49, "Environmental Qualification of Electrical Equipment Important to Safety of Nuclear Power Plants" and 10 CFR 50.65 "Requirements for Monitoring the Effectiveness of Maintenance at Nuclear Power Plants." Based upon these criteria, licensees' engineering organizations develop plant-specific listings of safety-related structures, systems, and components.

13.2.1   10 CFR Part 50, Appendix A, "General Design Criteria for Nuclear Power Plants"

Appendix A states the general requirements for establishing of QA controls. General Design Criterion I requires the following:

- Structures, systems, and components that are important to safety must be designed and fabricated to quality standards that are commensurate with the importance of the safety functions performed by the equipment.

- The licensee must implement an appropriate QA program to provide reasonable assurance that these structures, systems, and components will satisfactorily perform their safety functions.

- The licensee must maintain the QA program records of structures, systems, and components that are important to safety.

The scope of items that are "important to safety" includes a subset of plant equipment that is classified as "safety-related." QA program requirements for safety-related structures, systems, and components are contained in Appendix B to 10 CFR Part 50 (discussed below). Safety-related structures, systems, and components are defined in 10 CFR 50.2, "Definitions."

QA program controls that are appropriate for some types of non-safety-related equipment are contained in other regulatory guidance.

13.2.2   10 CFR Part 50, Appendix B, "Quality Assurance Criteria for Nuclear Power Plants and Fuel Reprocessing Plants"

Appendix B states the QA requirements that apply to activities that affect the safety-related functions of structures, systems, and components that prevent or mitigate the consequences of postulated accidents. The appendix defines quality assurance as all planned and systematic actions that are necessary to provide adequate confidence that structures, systems, and components will perform satisfactorily in service. Toward that end, Appendix B specifies 18 criteria that must be satisfied by the commitments in a licensee's QA program. These criteria cover such aspects as organizational independence, design control, procurement, document control, test control, corrective action, and audits. Meeting the criteria serves as a closed-loop management control process by which (1) planned controls ensure that safety-related activities are properly carried out, (2) verifications (inspections and audits) confirm that QA controls are effectively performed, (3) corrective actions are carried out when deficiencies are identified, and (4) records of QA aspects are maintained so that after-the-fact inspections can be performed (by the NRC), and documentation is maintained to demonstrate that the program was carried out properly.

Appendix B also stipulates that licensees establish measures to ensure that applicable regulatory requirements, design bases, and other requirements that are necessary to ensure adequate quality are suitably included or referenced in the documents for procurement of safety-related materials, equipment, and services whether purchased by the licensee or its contractors or subcontractors. Consistent with the importance and complexity of the products or services to be provided, licensees (or their designees) are responsible for periodically verifying that contractor's QA programs comply, as appropriate, with the applicable criteria in Appendix B and that they are effectively implemented.

The requirements of Appendix B are written at a high level, such that it was necessary for the NRC and the industry to develop consensus standards that give acceptable ways to conform to these requirements. The NRC then issued companion regulatory guides, which endorsed (with conditions, if warranted) QA codes and standards.

13.3 QA Regulatory Guidance

This section explains QA guidance for design and construction, operational activities, and staff reviews for licensing.

13.3.1 Guidance for Design and Construction Activities

In 1971, the American National Standards Institute (ANSI) issued the consensus QA Standard N45.2, "Quality Assurance Program Requirements for Nuclear Power Plants." The NRC endorsed ANSI N45.2 as providing a generally acceptable way to comply with the requirements of Appendix B to 10 CFR Part 50 during the design and construction phases of nuclear facilities. In later years, further ANSI standards promulgated in the N45.2 series, provided additional guidance on both programmatic aspects (such as records controls) and application-specific QA controls (such as those that apply to structures). The NRC conditionally endorsed these ANSI standards through its regulatory guides.

In 1979, the American Society of Mechanical Engineers (ASME) published NQA-1, "Quality Assurance Requirements for Nuclear Facilities," which consolidated eight of the ANSI QA standards regarding the programmatic aspects of Appendix B. The NRC endorsed the 1983 version of NQA-1 and the 1983-1a Addenda through Regulatory Guide 1.28, Revision 3, "Quality Assurance Program Requirements (Design and Construction)." Licensees with approved QA programs could voluntarily choose to adapt their programs to NQA-1. Two licensees amended their QA programs to adopt NQA-1; the remainder chose to remain committed to the ANSI series standards to meet Appendix B requirements.

13.3.2 Guidance for Operational Activities

In 1972, the American Nuclear Society issued N18.7, "Administrative Controls for Nuclear Power Plants," which focused on activities of importance for operating facilities, including reviews and audits, maintenance, tests, plant records management, and procedural controls. The NRC endorsed N18.7-1972 and N45.2-1971, in combination, as providing an acceptable way to satisfy the requirements of Appendix B to 10 CFR Part 50 for an operating facility. The dual endorsement caused some confusion in the industry. ANSI N18.7-1976 was then developed to better integrate the QA provisions from the various ANSI standards into a more cohesive framework. Specifically, ANSI N18.7-1976 referenced pertinent standards from the N45.2 series, and also extracted information from them to provide guidance on how operational phase activities of a similar nature should be carried out. For example, for design control of plant modifications, N18.7 states that the provisions of N45.2.11 shall be used. The NRC later conditionally endorsed the revised ANSI N18.7 standard in its Regulatory Guide 1.33, "Quality Assurance Program Requirements (Operations)," as complying with the requirements of Appendix B.

13.3.3 Guidance for Staff Reviews for Licensing

NUREG-0800, "Standard Review Plan for the Review of Safety Analysis Reports for Nuclear Power Plants," guides the staff review of applications. The staff uses the plan to initially review QA program descriptions in licensees' preliminary and final safety analysis reports, as well as program revisions during facility operation. The Plan has specific review guidance correlated with the 18 criteria of Appendix B to 10 CFR Part 50, and integrates a review of licensee commitments to adopt the NRC's QA-related regulatory guides and the industry's QA codes and standards.

13.4 QA Programs

This section discusses programs conducted under the Revised Reactor Oversight Process, augmented programs, and graded QA.

13.4.1 Background

The NRC previously relied upon inspections that focused on licensee-generated QA documentation (such as procedures and records) to establish whether licensees had properly carried out their QA commitments for nuclear facilities. This methodology emphasized examining licensees' programmatic QA activities, and verifying documentation (such as quality control inspection reports and audit reports) and procedures. In the late 1980s, the NRC shifted the focus of its inspections to observation of in-process work activities, and independent confirmation of the quality of hardware at nuclear facilities. This approach emphasized the determination of the licensees' attainment of quality by observing plant staff actually performing work activities. When detecting problems during the work observations, the NRC inspector was to ascertain which aspects of the QA program were not carried out properly and, thus, led to the problem. The licensee then carried out corrective actions to properly address the identified problem.

13.4.2 QA Under the Revised Reactor Oversight Process

In April 2000, the NRC implemented its revised Reactor Oversight Process for operating reactors, of which baseline inspections are a component. (See Article 6.) Under the baseline inspection program, there is one primary procedure related to QA issues, which is known as Inspection Procedure 71152, "Identification and Resolution of Problems." Inspectors use this procedure to assess the effectiveness of licensees' programs to identify and resolve problems according to a performance-based review of specific issues. In particular, inspectors look for cases in which a licensee may have missed generic implications of specific problems, and for the risk significance of combinations of problems that individually may not have significance. They also verify that licensees are properly capturing issues that could affect the availability of equipment that is tracked under 10 CFR 50.65, "the Maintenance Rule," or by performance indicators. They do not inspect other aspects of QA program implementation in the baseline inspection program, but may, through supplemental inspections.

13.4.3 Augmented Quality Programs

Some equipment in the nuclear facility may be classified as non-safety-related, and yet still be important to safety for some unique reason. In specific cases, the NRC has specified that QA controls are warranted for equipment determined to be more important than commercial-grade equipment. However, the QA controls would not have to meet Appendix B requirements, which apply only to activities affecting safety-related functions. Typically, applying QA controls to this important-to-safety, yet non-safety-related, equipment is called "augmented quality control." Specific examples include equipment that applies to fire protection, station blackout, and anticipated transients without scram. Branch Technical Position CMEB 9.5-1, "Guidelines for Fire Protection for Nuclear Power Plants," gives 10 quality program elements that should be applied to the design, procurement, installation, and testing of fire protection systems in safety-related areas of the plant. Regulatory Guide 1.155, "Station Blackout," gives 10 quality elements that apply to non-safety-related equipment that is required to meet the Station Blackout Rule (10 CFR 50.63). Similarly, NRC Generic Letter 85-06, "Quality Assurance Guidance for Anticipated Transient Without Scram Equipment That Is Not Safety-Related," defines less-stringent QA controls for such equipment to meet 10 CFR 50.62, "Requirements for Reduction of Risk from Anticipated Transient Without Scram Events for Light-Water cooled Nuclear Power Plants."

13.4.4 Graded QA

In 1993, the NRC embarked on an effort with the industry to develop guidance for graded QA. The goal was to formulate guidance on how Appendix B QA controls could be graded by a process that included the following four essential elements:

(1) a process that, with high confidence, will identify the proper safety significance of all structures, systems, and components in a reasonable and consistent manner

(2) an effective root-cause analysis and corrective action program

(3) the determination of appropriate QA controls for individual or groups of structures, systems, and components, consistent with their safety function and safety significance

(4) a way of reassessing the safety significance of systems, structures, and components and related QA controls as new information becomes available

In parallel with the volunteer project efforts, the staff continued developing a regulatory guide on graded QA (DG-1064, "An Approach for Plant-Specific, Risk-Informed Decisionmaking: Graded Quality Assurance"). The staff issued this guidance for public comment in 1997 along with a set of companion regulatory guides and standard review plan sections for risk-informed regulatory decision-making. It issued the final guidance on graded quality assurance as Regulatory Guide 1.176, "An Approach for Plant-Specific, Risk-Informed Decision Making: Graded Quality Assurance," in 1998. (See Article 10, Section 10.3.6 of this report for more details.) The NRC is not planning to revise Regulatory Guide 1.176 until it gains further experience from other efforts and the below-noted rule change.

The NRC staff encountered significant challenges while developing the graded quality assurance approach described in Regulatory Guide 1.176. Some of these challenges included:

- Establishing a baseline for the quality, scope, and rigor of the Probabilistic Risk Assessment upon which the determination of safety significance of plant equipment would be based

- Formulating a realistic quantitative estimate of the change in plant risk arising from implementing graded quality assurance

- Defining the expert panel composition, roles, and responsibilities

- Establishing the level of documentation and detail needed to describe expert panel deliberations and to support any required corrective actions

- Defining the scope, form, and content of the requisite performance monitoring program that would be used to judge the success and effectiveness of the graded quality assurance program

- Ascertaining the effect of graded quality assurance program on commitments to industry standards relied upon by licensees to satisfy Appendix B to 10 CFR Part 50 requirements

- Formulating the extent of and need for NRC staff-industry regulatory interface during implementation

- Determining the effect on graded quality assurance of 10 CFR 50.54 rulemaking (described below)

13.4.5 Other Quality Assurance-Related Activities

The staff issued a rule change on April 26, 1999, for paragraph (a) of 10 CFR 50.54, "Conditions of Licenses." This change allows licensees to apply any alternatives or exceptions to a QA program that have been approved by the NRC for use by another facility, without prior NRC approval, if the bases for the approval can be shown to apply to the plant making the change.

ARTICLE 14. ASSESSMENT AND VERIFICATION OF SAFETY

Each Contracting Party shall take the appropriate steps to ensure that:

(i) comprehensive and systematic safety assessments are carried out before the construction and commissioning of a nuclear installation and throughout its life. Such assessments shall be well documented, subsequently updated in the light of operating experience and significant new safety information, and reviewed under the authority of the regulatory body;

(ii) verification by analysis, surveillance, testing, and inspection is carried out to ensure that the physical state and the operation of nuclear installations continues to be in assurance with its design, applicable national safety requirements, and operational limits and conditions.

This section explains the governing documents and process for ensuring that systematic safety assessments are carried out during the life of the nuclear installation, including for the period of extended operation. It focuses on assessments performed to maintain the licensing basis of a nuclear installation. It also discusses experience and lessons-learned from performing safety assessments. Finally, this section explains verification of the physical state and operation of the nuclear installation by analysis, surveillance, testing and inspection.

Other articles, for example, Articles 6, 10, 13, 18, and 19, also discuss activities undertaken to achieve nuclear safety at nuclear installations.

14.1 Ensuring Safety Assessments Throughout Plant Life

Before a nuclear facility is constructed, commissioned, and licensed, an applicant must perform comprehensive and systematic safety assessments, which are reviewed and approved by the NRC. These assessments and reviews are discussed in Article 18.

This section focuses on the assessments that are required throughout the life of a nuclear installation; that is, assessments required to maintain the licensing basis. To show conformance with the licensing basis, a licensee must maintain records of the original design bases and any changes. This section explains how such changes are documented, updated and reviewed. Further, renewal of a license is predicated on the requirement that a licensee will continue to meet its current licensing basis, and this section explains how this requirement is accounted for in license renewal.

14.1.1 Background

10 CFR 50.2, "Definitions," defines design bases as "that information which identifies the specific functions to be performed by a structure, system, or component of a facility, and the specific values or ranges of values chosen for controlling parameters as reference bounds for design."

Although the terms "current licensing basis" and "licensing basis" are widely used in matters related to power reactors operating in accordance with the regulations in 10 CFR Part 50, the terms are not defined in Part 50 or major regulatory guidance related to Part 50. 10 CFR Part 54.3, "Definitions," defines "current licensing basis" pertaining to license renewal for nuclear installations as follows:

> Current licensing basis is the set of NRC requirements applicable to a specific plant and a licensee's written commitments for ensuring compliance with and operation within applicable NRC requirements and the plant-specific design basis (including all modifications and additions to such commitments over the life of the license) that are docketed and in effect. The current licensing basis includes the NRC regulations contained in 10 CFR Parts 2, 19, 20, 21, 26, 30, 40, 50, 51, 54, 55, 70, 72, 73, 100 and appendices thereto; orders; license conditions; exemptions; and technical specifications. It also includes the plant-specific design-basis information defined in 10 CFR 50.2 as documented in the most recent final safety analysis report as required by 10 CFR 50.71 and the licensee's commitments remaining in effect that were made in docketed licensing correspondence such as licensee responses to NRC bulletins, generic letters, and enforcement actions, as well as licensee commitments documented in NRC safety evaluations or licensee event reports.

A definition of the "licensing basis" and a detailed description of its major elements is provided by the Office of Nuclear Reactor Regulation Office Instruction LIC-100, "Control of Licensing Bases for Operating Reactors," dated March 2001.

The NRC carries out regulatory programs to give reasonable assurance that plants continue to conform to the licensing basis. (Article 6 discusses these programs.)

14.1.2 Maintaining the Licensing Basis

This section explains the governing documents and process used to maintain the licensing basis. The main governing documents are "10 CFR 50.90, "Application for Amendment of License or Construction Permit," 10 CFR 50.59, "Changes, Tests, and Experiments, and 10 CFR 50.71, "Requirements for Updating of Final Safety Analysis Reports." The section also discusses relevant experience and examples.

14.1.2.1 Governing Documents and Process

A licensee is to operate its facility in accordance with the license, and as described in its final safety analysis report. To change its license or reactor facility, a licensee must follow the review and approval processes established in the regulations. For license amendments, including changes to technical specifications, the licensee must request NRC approval in accordance with 10 CFR 50.90, "Application for Amendment of License or Construction Permit." However, licensees can make certain changes without prior NRC approval if they perform specified reviews and meet certain conditions. Such changes are provided for in 10 CFR 50.59, "Changes, Tests, and Experiments," as described below.

10 CFR 50.59, "Changes, Tests, and Experiments"

10 CFR 50.59 describes the circumstances under which a licensee may make changes to its facility or procedures as described in its final safety analysis report and may conduct tests and experiments that are not described in its report without prior NRC approval. An example of such a circumstance is if the change does not affect the technical specifications. Licensees are required to periodically submit information about changes made in accordance with 10 CFR 50.59. The NRC monitors each licensee's processes for implementing the requirements in 10 CFR 50.59.

NRC approval is required in circumstances when the change, test, or experiment would result in more than a minimal increase in the frequency of occurrence of an accident previously evaluated in the final safety analysis report. The licensee must apply to amend the license pursuant to 10 CFR 50.90 under such circumstances. The NRC performs and documents a safety evaluation in these instances before it authorizes the change.

In 1995, the NRC recognized that Millstone Unit 1 had conducted refueling outages in a manner outside its design bases, as shown by the analysis and assumptions in its updated safety analysis report. This recognition led to questions about the regulatory framework that authorizes licensees to make changes to their facilities without prior NRC approval. As a result, the staff initiated a review of the 10 CFR 50.59 process to identify short- and long-term actions to improve implementation and oversight of the process. In summary, the staff found that difficulties in day-to-day use arise when the meaning of the language of the rule is not clear, leading the staff and licensees to interpret the rule differently and have different expectations about its implementation.

As a result of the issues raised about 10 CFR 50.59, the NRC conducted a rulemaking to clarify the requirements and to provide a limited degree of flexibility for licensees to make certain changes that only "minimally" increase the probability or consequences of accidents. The NRC issued the final rule on October 4, 1999; it took effect on March 13, 2001. In November 2000, the NRC issued a corresponding Regulatory Guide 1.187, "Guidance for Implementation of 10 CFR 50.59, Changes, Tests, and Experiments." This guide endorses an industry guidance document, NEI 96-07, Revision 1, " Guidelines for 10 CFR 50.59 Implementation," also issued in November 2000.

To help licensees (and the NRC staff) determine which information is relevant to the "design bases" as defined in Section 50.2, the NRC issued Regulatory Guide 1.186, "Guidance and Examples of Identifying 10 CFR 50.2 Design Bases," in December 2000. This guide endorses Appendix B to NEI 97-04,"Nuclear Energy Institute Design Bases Program Guidelines," issued in September 1997, as providing general guidance and examples that are acceptable for identifying information on design bases.

10 CFR 50.71 Requirements for Updating of Final Safety Analysis Reports

Another process for making changes is set forth in Paragraph (e) of 10 CFR 50.71, "Requirements for Updating of Final Safety Analysis Reports," which requires licensees to update their final safety analysis reports periodically to incorporate the information and analyses that they submitted to the Commission or prepared pursuant to Commission requirements. Revisions to the updated final safety analysis reports are to include the effects of changes that occur in the vicinity of the plant, changes made in the facility or procedures described in the report, safety evaluations for approved license amendments and for changes made under 10 CFR 50.59, and safety analyses conducted at the request of the Commission to address new safety issues.

14.1.2.2 Experience and Examples

In 1997 and 1998, the NRC conducted architect/engineer design inspections at a number of plants. Some of the issues raised during inspections at the DC Cook Nuclear Plant contributed to its long shutdown and precluded restart until the licensee took appropriate corrective actions to restore conformance of the facility to its design bases. Given the overall results of these inspections, the NRC concluded that the agency did not need to expend this level of effort for such inspections, but that it was appropriate to maintain NRC inspection oversight of licensees' design and configuration control processes. As discussed in Article 6, the NRC completely revised its inspection program for operating power reactors. However, it retains an inspection procedure in the baseline program for reviewing safety system design and performance capability and licensee processes to modify the plant.

14.1.3 License Renewal

This section explains license renewal. It covers the governing documents and regulatory process, recent experience and examples.

14.1.3.1 Governing Documents and Process

10 CFR Part 54, "Requirements for Renewal of Operating Licenses for Nuclear Power Plants"

The Atomic Energy Act and NRC regulations limit commercial power reactor licenses to 40 years, but also permit such licenses to be renewed. The original 40-year term was selected on the basis of economic and antitrust considerations, rather than by technical limitations. However, the selection of this term may have resulted in individual plants being designed on the basis of an expected 40-year service life.

10 CFR Part 54, known as the "license renewal rule," establishes the technical and procedural requirements for renewing operating licenses. License renewal requirements for power reactors are based on two key principles:

(1) The regulatory process, which assesses and verifies safety, continued into the extended period of operation, is adequate to ensure that the licensing basis of all currently operating plants provides an acceptable level of safety. The possible exception is detrimental effects of aging on certain systems, structures, and components, and possibly a few other issues applying to safety only during the period of extended operation, and

(2) Each plant's licensing basis is required to be maintained throughout the renewal term.

The foundation of license renewal rests on the determination that currently operating plants continue to maintain an adequate level of safety. Over the plant's life, this level has been enhanced by maintaining the licensing basis, properly adjusted to incorporate new information that is derived from operating experience. Moreover, NRC activities have continually ensured that the licensing basis will continue to provide an acceptable level of safety.

The NRC has developed guidance documents for license renewal. It developed Regulatory Guide 1.188, "Standard Format and Content for Applications to Renew Nuclear Power Plant Operating Licenses," to guide applicants in applying to renew a license and the "Standard Review Plan for the Review of License Renewal Applications for Nuclear Power Plants," NUREG-1800, to guide the staff in reviewing applications. The standard review plan references the "Generic Aging Lessons Report," NUREG-1801, which generically documents the basis for determining when existing programs are adequate, and when they should be augmented for license renewal. The NRC issued all of these documents in the summer of 2001.

10 CFR Part 54 requires applicants to identify all plant systems, structures, and components that are within the scope of renewal. Specifically, these are (1) all safety-related systems, structures, and components; (2) all systems, structures, and components, the failure of which could prevent the accomplishment of safety-related functions; and (3) systems, structures, and components that are relied on to demonstrate compliance with the NRC's regulations for fire protection (10 CFR 50.48), environmental qualification (10 CFR 50.49), pressurized thermal shock (10 CFR 50.61), anticipated transients without scram (10 CFR 50.62), and station blackout (10 CFR 50.63).

Under the license renewal rule, the applicant must perform a screening review of all systems, structures, and components within the scope of the rule to identify "passive" and "long-lived" structures and components. The applicant must demonstrate that it will manage the effects of aging such that the structures, systems, and components will function as intended throughout the period of extended operation. Active equipment is considered to be adequately monitored under the current regulatory process because any detrimental aging effects are readily detectable, and will be identified and corrected by routine surveillances and performance testing.

The rule also requires applicants to identify and update time-limited aging analyses for systems, structures, and components that may be based on the length of the current operating license term. For example, the analyses supporting environmental qualification of electric equipment explicitly consider the current 40-year operating license term and must be reevaluated for license renewal.

Additionally, the NRC's environmental protection regulation, 10 CFR Part 51, "Environmental Protection Regulations for Domestic Licensing and Related Regulatory Functions," applies to license renewal. The NRC amended this regulation to enhance the agency's environmental review process for license renewal. Specifically, the review requirements for 10 CFR Part 51 are founded on the conclusion that certain environmental issues can be resolved generically and need not be evaluated in each plant-specific application. These issues are described in NUREG-1437, "Generic Environmental Impact Statement for License Renewal of Nuclear Plants." Supplement 1 to NRC Regulatory Guide 4.2, "Preparation of Supplemental Environmental Reports for Applications to Renew Nuclear Power Plant Operating Licenses," guides applicants in preparing environmental reports for license renewal. The "Standard Review Plan for Environmental Reviews for Nuclear Power Plants, Operating License Renewal," NUREG-1555, Supplement 1, guides the NRC staff's review of the environmental issues of an application for renewal.

The decision regarding whether to seek license renewal rests with the licensees, who must make business decisions as to whether they are likely to satisfy the relevant NRC requirements and must evaluate the costs of the venture. The NRC's role is to establish a reasonable and stable review process and such safety and environmental protection standards that licensees can make timely decisions about whether to seek license renewal.

14.1.3.2 Experience and Examples

The NRC reviewed the first license renewal applications for the Calvert Cliffs Nuclear Power Plant and the Oconee Nuclear Station and issued the renewed licenses in 2000. The agency also completed its review of the Arkansas Nuclear One, Unit 1, application and issued the renewed license in June 2001. The NRC is now reviewing license renewal applications for Hatch, Turkey Point, Surry, North Anna, McGuire, Catawba and Peach Bottom. The NRC has committed to complete the reviews within 30 months of receipt if a hearing is conducted, and within 25 months if not. Almost all other licensees are expected to seek renewal of their licenses in the future.

14.2 Verification by Analysis, Surveillance, Testing and Inspection

Licensees are required to verify that they are operating their nuclear installations in accordance with the plant-specific design and requirements. Requirements specifying verification are contained in the Technical Specifications (for surveillance) and national consensus codes (for testing and periodic inspections).

10 CFR 50.55a, "Codes and Standards," defines the requirements for applying industry codes and standards to nuclear power reactors during design, construction, and operation. This regulation applies to both operating licenses and construction permits. 10 CFR 50.55a states, "Systems and components of boiling and pressurized water-cooled nuclear power reactors must meet the requirements of the ASME Boiler and Pressure Vessel Code specified in paragraphs (b) though (g) of this section." The ASME Code has requirements for the construction and periodic inspection of boilers, pressure vessels, and nuclear components. These requirements apply to materials, design, fabrication, testing, inspection, and stamping. 10 CFR 50.55a also provides for alternatives to the ASME Code when authorized by the NRC.

Through analysis, surveillance, testing, and inspection, the NRC verifies that the physical state and operation of nuclear installations continue to be in accordance with the designs, applicable national safety requirements, and operational limits and conditions. As previously discussed in Article 6, NRC's reactor oversight process includes inspections to verify that licensees are fulfilling their obligations to conduct such surveillances and testing and take corrective action. The NRC staff updates, revises, and improves existing regulatory programs in light of operating experience and significant new safety information. These activities are discussed in Article 19.

ARTICLE 15. RADIATION PROTECTION

Each Contracting Party shall take the appropriate steps to ensure that, in all operational states, the radiation exposure to the workers and to the public caused by a nuclear installation shall be kept as low as is reasonably achievable, and that no individual shall be exposed to radiation doses that exceed the prescribed national dose limits.

This section summarizes the authorities and principles of radiation protection, the regulatory framework, regulations, and radiation protection programs for controlling radiation exposure for occupational workers and members of the public.

Article 17 addresses radiological assessments that apply to licensing and to facility changes.

15.1 Authorities and Principles

Generally, U.S. radiation control measures are founded on radiological risk assessments by the United Nations Scientific Committee on the Effects of Atomic Radiation (UNSCEAR) and the U.S. National Academy of Sciences Committee on the Biological Effects of Ionizing Radiation (BEIR). These risk assessments are reflected in the risk management recommendations promulgated by the International Commission on Radiological Protection (ICRP) and the National Council on Radiation Protection and Measurements (NCRP). On the basis of these assessments and recommendations, the Environmental Protection Agency (EPA) develops "generally applicable radiation standards" for use by the other Federal agencies, including the NRC. Considering these recommendations and standards, the responsible agencies, such as the NRC, then establish regulations.

The principles upon which the U.S. radiation protection programs are based are generally consistent with the principles espoused by the ICRP. That is to say, (1) it is known that large doses of ionizing radiation can be deleterious to human health, and (2) it is considered prudent to assume that small doses may also be harmful, with the probability of a deleterious effect being proportional to the dose. The ICRP-recommended protection principles of "limitation," "justification," and "optimization" are acknowledged, but are proving difficult to carry out.

Of these, "limitation" is most practicable. Dose limits are established in the regulations, and these limits cannot be exceeded without violating the regulations. There is a lengthy history of the doses being kept within the limits for workers (NUREG-0713, Vol. 21, 1999) and members of the public living near nuclear power plants (NUREG/CR-2850, Vol. 14, 1996).

"Justification," the recommendation that any activity involving radiation exposure should be shown to be beneficial before the activity is undertaken, has proved impracticable. The difficulty results from three basic facts. Specifically, (1) every human activity involves radiation exposure, (2) the outcome of a new activity can never be determined in advance, and (3) the U.S. Government (like other governments) lacks this degree of control over the activities of its citizens. Thus, the "justification" activities in the U.S. are generally limited to cost/benefit studies and analyses of the environmental impact of major actions, such as imposing a new regulation or building a new nuclear power plant.

Rather than "optimization," the U.S. has used the expression "as low as is reasonably achievable" (ALARA), although the two concepts are consistent. As a guiding principle, ALARA (with varying terminology) dates back to 1939, at least in the U.S., and ALARA is defined in the regulations for occupational workers and for members of the public.

For decades, the ALARA criterion for occupational radiation exposure has been addressed in 10 CFR Part 20, "Standards for Protection Against Radiation," but as an admonition rather than requirement. In 1994, the regulation was changed to require that all licensees develop, document, and carry out an ALARA Program. Compliance with this requirement was to be judged on the basis of a licensee's capability to track and, if necessary, reduce exposures, and not on whether exposures and doses represented an absolute minimum or whether the licensee had used all possible methods to reduce exposures.

For control of radiation exposure to members of the public, the NRC modified 10 CFR Part 50 by adding Appendix I, "Numerical Guides for Design Objectives and Limiting Conditions for Operation to Meet the Criterion As Low As Is Reasonably Achievable for Radioactive Material in Light-Water-Cooled Nuclear Power Reactor Effluents." Issued in 1975, this appendix required that radioactive releases from nuclear power plants be kept ALARA. This requirement led to the establishment of numerical criteria (i.e., 0.00005 Sv (0.005 rem)) in a year to the most highly exposed individual). This NRC requirement was soon followed by similar EPA requirements for other facilities. It is not clear that these requirements satisfy the ICRP's intent, but they are sufficient to keep public doses well below the local variation in doses from natural sources.

Although U.S. regulations are generally consistent with ICRP recommendations, to date, certain constraints have limited the extent to which the U.S. regulations coincide with the ICRP recommendations. One important constraint has been the desire for regulatory stability. Revising the regulations to incorporate every new ICRP position would impose a serious burden on the licensees without a commensurate benefit. Furthermore, for nuclear power reactors, new requirements are constrained by the "Backfit Rule" (10 CFR 50.109), which essentially requires that any increase in regulatory requirements be justified by a commensurate improvement in safety. Consequently, U.S. regulations were founded on older (rather than the most recent) recommendations of the ICRP. Nevertheless, the Commission may consider and then elect to implement a new international standard (e.g., ICRP-60).

15.2 Regulatory Framework

Requirements for radiation protection were developed to implement laws passed by Congress. These laws are the Atomic Energy Act of 1954, the Energy Reorganization Act of 1974, and the Uranium Mill Tailings Radiation Control Act of 1978.

The direct controls over licensees are established primarily by NRC regulations. However, the regulations lack the specificity and detail that are needed to achieve the NRC's objectives. Various documents provide additional guidance and clarification. Specifically, these documents include regulatory guides, topical staff and contractor reports (NUREG series), generic letters, technical specifications, and license conditions. These documents are supported by international standards, consensus national standards, and authoritative recommendations (such as those of the ICRP and NCRP). However, these supporting documents have no official status unless they are referenced in or adopted by a regulation or documents providing regulatory guidance, such as regulatory guides or standard review plans. Of particular importance are NUREG-0800, the "Standard Review Plan," which guides the staff in reviewing safety analysis reports and Regulatory Guide 1.70, "Standard Format and Content of Safety Analysis Reports," which guides the applicant in writing safety analyses. Chapter 11, "Radioactive Waste Management," of the Standard Review Plan addresses the control of radioactive effluents. Chapter 12 addresses "Radiation Protection." Chapter 15, "Accident Analysis," details how to calculate offsite and control room operator doses for design-basis accidents. Paragraph (g) of 10 CFR 50.34, "Conformance with the Standard Review Plan," makes the evaluation of the facility against the Standard Review Plan a requirement.

As discussed under Article 6, the Revised Reactor Oversight Process has cornerstones for radiation safety. The cornerstone, "Public Radiation Safety," focuses on the effectiveness of the plant's programs to meet applicable Federal limits involving the exposure, or potential exposure, of members of the public to radiation and ensure that the effluent releases from the plant are ALARA. The cornerstone, "Occupational Radiation Safety," focuses on the effectiveness of the plant's program(s) to maintain worker dose within the regulatory limits and provide occupational exposures that are ALARA.

15.3 Regulations

This section summarizes the provisions of the regulations that apply to radiation protection. These regulations are 10 CFR Part 20, "Standards for Protection Against Radiation" and 10 CFR Part 50, "Domestic Licensing of Production and Utilization Facilities."

15.3.1 10 CFR Part 20, "Standards for Protection Against Radiation"

This part of the NRC regulations establishes requirements for radiation protection for all NRC licensees. The requirements in 10 CFR Part 20 are supplemented by specific requirements for specific operations and specific kinds of licenses. In particular, these supplementary requirements include Part 30 for byproduct material licensees, Part 34 for radiographic operations, Part 35 for medical users of byproduct material, Part 39 for oil well logging, Part 40 for users of source materials, Part 50 for nuclear power plants, Part 70 for special nuclear material users, and Part 71 for the transport of radioactive materials.

Although amended many times, the basic requirements of 10 CFR Part 20 have remained in effect for more than three decades. The most recent major revision was issued in 1991, and became fully effective in 1994. Fundamental changes in this revision included combining internal and external doses under the same limit, and reducing the basic dose limit to 0.05 Sv (5 rem) annually, with no provision for permitting higher doses on the basis of worker age. Other general changes included switching the emphasis from material control requirements to dose limits, and considering internal and external doses as having equal importance. The 1991 revision was largely predicated on the recommendations in ICRP-26, "Recommendations of the International Commission on Radiological Protection (Adopted January 17, 1977)," issued in 1991, ICRP-30, "Limits of Intakes of Radionuclides by Workers," 1978-1982, and NCRP Report No. 91, "Recommendations On Limits For Exposure to Ionizing Radiation," issued in June 1987.

10 CFR Part 20 provides relatively comprehensive coverage of general requirements for radiation protection. Subpart A, "General Provisions," defines dose units, most of which are the

same as those defined by the ICRP. However, the unit identified as "deep dose equivalent" is specific to the United States. The deep dose equivalent closely corresponds to the ICRP's "ambient dose equivalent" [H*(d)], where the depth (d) is 1 cm. 10 CFR Part 20 also defines the total effective dose equivalent (TEDE) as the sum of the deep dose equivalent for external exposures, and the committed effective dose equivalent for internal exposures.

10 CFR Part 20, Subpart B, "Radiation Protection Programs," requires and describes a radiation protection program. The description and discussion are quite general, however, because they apply to a wide range of licensees, ranging from a small instrument shop (with a few sealed sources) to a nuclear fuel reprocessing plant.

10 CFR Part 20, Subpart C, "Occupational Dose Limits," establishes annual dose limits for adults of 0.05 Sv (5 rem) TEDE, 0.15 Sv (15 rem) to the lens of the eye, 0.5 Sv (50 rem) committed dose equivalent to any other internal organ or tissue, and 0.5 Sv (50 rem) shallow dose equivalent to the skin or any extremity. In addition to the basic limits, there is a provision for exceeding these limits in "planned special exposures." The conditions required for a planned special exposure are quite stringent, so the use of this provision is considered highly unlikely. The limits for people under 18 years of age are 10 percent of the limits for people over 18 years of age. The dose limit for the embryo/fetus of a *declared* pregnant woman is .005 mSv (0.5 rem) during the entire pregnancy. (This limit applies to any woman who formally declares herself pregnant, and does not apply to any woman who does not make such a formal declaration.)

10 CFR Part 20, Subpart D, "Radiation Dose Limits for Individual Members of the Public," establishes a general limit of .001 Sv (0.1 rem), excluding doses from nature and medical procedures. In special cases, the NRC may permit higher doses or impose lower limits.

10 CFR Part 20, Subpart E, "Radiological Criteria for License Termination," revised on July 21, 1997, addresses radiological criteria for license termination. Specifically, this regulation states that a site will be considered acceptable for unrestricted use if the annual dose to the average member of the critical group would not exceed .00025 mSv (.025 rem) TEDE. (A critical group is defined as the group of individuals who are reasonably expected to receive the greatest exposure to residual radioactivity for any applicable set of circumstances. The residual radioactivity must have been reduced to a level that is ALARA). This regulation also states that a site is considered acceptable for license termination under restricted conditions if the licensee has enforceable controls that provide reasonable assurance that the annual dose would not exceed .00025 Sv (.025 rem) TEDE. Without the enforceable controls, the stated limit is .001 Sv (0.1 rem) TEDE, subject to other specified conditions. The licensee is further required to demonstrate that these levels of residual radioactivity are ALARA.

10 CFR Part 20, Subpart F, "Surveys and Monitoring," requires personal monitoring for people who are likely to receive doses in excess of 10 percent of an applicable limit. Subpart F also requires periodic surveys of work areas and equipment.

10 CFR Part 20, Subpart G, "Control of Exposure from Extended Sources in Restricted Areas," specifies requirements for controlling access to a "high radiation area," where the dose may exceed .001 Sv (0.1 rem) in an hour at any point that is 30 cm or more from a radiation source, or to a "very high radiation area," where the radiation level could be 500 rads (5 grays) or more in an hour at 1m or more from a source.

10 CFR Part 20, Subpart H, "Respiratory Protection and Controls to Restrict Internal Exposure in Restricted Areas," discourages relying on personal respiratory protection, but permits using such equipment if appropriate requirements are met, such as certification, testing, written procedures, training, and a medical examination. Appendix A to Part 20 identifies specific protection factors for the various respirators. Appendix B to Part 20 lists the nuclide-specific "annual limits on intake" and "derived air concentrations".

10 CFR Part 20, Subpart I, "Storage and Control of Licensed Material," requires security of stored material and control of material that is not in storage.

10 CFR Part 20, Subpart J, "Precautionary Measures," stipulates requirements for posting and labeling, as well as exemptions from those requirements. It also specifies procedures for receiving and opening packages. Appendix A to Part 20 states the quantities of the various radionuclides that require labels.

10 CFR Part 20, Subpart K, "Waste Disposal," details methods for obtaining approval for disposal, release to sanitary sewerage, obtaining approval for disposal by incineration, and disposal by transfer to disposal sites. Appendix F to Part 20 states detailed requirements for transfer to disposal sites. Furthermore, this subpart provides for disposal of specified wastes as though they were nonradioactive. Specifically, these wastes are liquid scintillation fluid or animal tissue that contains no more than 1.85 kBq (0.05 microcuries) per gram of tritium or carbon-14.

10 CFR Part 20, Subpart L, "Records," requires that records be kept on radiation and contamination surveys, individual monitoring, planned special exposures, doses to members of the public, and waste disposal.

10 CFR Part 20, Subpart M, "Reports," requires reports to the NRC by telephone or in writing (or both) for a variety of events, including the theft or loss of licensed material, incidents in which specified dose limits may be exceeded, actual exposures or concentrations in excess of the limits, and planned special exposures. Subpart M also requires licensees to prepare annual reports on individual occupational doses to workers in nuclear power plants, fuel cycle facilities, or facilities that are licensed to possess quantities of byproduct material that exceed certain amounts.

10 CFR Part 20, Subpart N, "Exemptions and Additional Requirements," specifies that the NRC may grant exemptions from regulatory requirements or impose additional requirements.

10 CFR Part 20, Subpart O, "Enforcement," enables the NRC to obtain injunctions to prevent violations, impose civil penalties for violations, or revoke a license for a violation. Subpart O also provides criminal penalties for willful or attempted violations, or for conspiracy to violate a regulation.

15.3.2 10 CFR Part 50, "Domestic Licensing of Production and Utilization Facilities"

10 CFR Part 50 is the principal regulation that addresses the safety of nuclear power plants. However, only a small part directly addresses radiation protection. (The revised dose criteria for design-basis accidents are stated in 10 CFR 50.34(a)(1)(ii)(D) for future licensing actions after implementation of the revised rule in 1997. The dose criteria for most operating nuclear power reactors are stated in 10 CFR 100.11(a).) Even so, the parts of 10 CFR Part 50 that do affect radiation protection are significant. Of particular importance are paragraph (a) of 10 CFR 50.34, "Design Objectives for Equipment to Control Releases of Radioactive Material in Effluents — Nuclear Power Reactors," Appendix I to 10 CFR Part 50, and paragraph (g) of 10 CFR 50.34, "Conformance with the Standard Review Plan," which requires NRC review of the in-plant radiation protection program.

15.4 Radiation Protection Activities

This section discusses radiation protection activities that apply to occupational workers and to members of the public.

15.4.1 Control of Radiation Exposure of Occupational Workers

Occupational radiation exposure control was initially challenged by efforts to minimize costs by using as few workers as practicable, and workers with the minimum qualifications needed to perform work activities. In 1971, the relevant ANSI standard for nuclear power plants (ANSI-18.1-1971) endorsed a nuclear power plant staff without even a single position occupied by a person with a college degree, and the radiation protection staff consisted of a technician working half-time (the other half of the time, this person was assigned to the chemistry staff). This standard was adopted by Regulatory Guide 1.8, "Personnel Selection and Training," issued in 1971. The objective was for the technical problems to be solved by the nuclear steam system supplier and the architect/engineer. However, as soon as a few commercial nuclear power plants went into operation, it became clear that this objective had not been achieved.

As a result, the NRC issued Regulatory Guide 8.8, "Information Relevant to Maintaining Occupational Radiation Exposure As Low As Practicable (Nuclear Reactors)," in 1973. This guide called for extensive efforts to keep doses ALARA, the participation of qualified health physicists in the design phase, and the involvement in the operational phase of a health physics manager whose qualifications were equivalent to those required for certification by the American Board of Health Physics. Although the qualification requirement was subsequently reduced, the need for health physics expertise had been established. In 1977, the NRC revised Regulatory Guide 1.8 "Personnel Selection and Training," to address the need for health physics expertise.

The foregoing comments do not imply that the NRC's oversight and regulation of the radiation protection programs is limited to focusing on personnel qualifications. For licensing, the NRC's review is comprehensive. As a result, it ensures that the safety analysis report and radiation protection plan properly address each item in 10 CFR Part 20, as well as the "Instruction to Workers" provisions of 10 CFR Part 19 and the provisions of the relevant regulatory guides, (such as 1.8, 8.8, and 8.10) and NUREG-0761, "Contents of Radiation Protection Plans for Nuclear Power Reactor Licensees."

Once the NRC issues a license, it maintains an active regulatory program, which includes daily monitoring of licensee and regional reports to alert the NRC staff of potential problems in radiation safety. These problems range from major repairs of highly radioactive components inside the facility to contamination from small leaks of liquid and gaseous materials. The staff evaluates the reports and discusses them with regional NRC inspection staff. Significant health physics problems can trigger significant reactive regional inspections or a generic communication to the industry.

The program for occupational radiation control has succeeded in reducing doses. The NRC staff has been collecting the annual occupational exposure data for light-water reactors since 1969. The doses are strongly influenced by the amount and kind of maintenance performed, so the data fluctuate from year to year. Still, clear trends are evident. Using the average collective dose per reactor as the reference statistic, one can conclude that the doses were almost randomly variable before the accident at Three Mile Island Unit 2. Thereafter, the doses increased as a result of the extensive modifications required of all nuclear power plants in response to new requirements. The average dose reached a peak of 7.91 person-Sv (791 person-rem) per reactor in 1980. Since then, doses have declined almost steadily to 1.31 person-Sv (131 person-rem) per reactor in 1999 (the last year for which the data have been compiled). The 1999 value of 1.31 person-Sv (131 person-rem) per reactor was the second lowest average collective dose recorded since data collection began in 1969. The lowest recorded average collective dose of 1.26 person-Sv (126 person-rem) per reactor occurred in 1998. Between 1997 and 1999, the average collective dose for boiling-water reactors exceeded the average collective dose for pressurized-water reactors by 75%. In 1999, 75,420 workers at nuclear plants received 137 person-Sv (13,666 person-rem) for an average of .0018 Sv (0.18 rem) per worker. This average worker dose has been steadily declining from an average dose per worker of .0075 person-Sv (0.75 person-rem) in 1974.

15.4.2 Control of Radiation Exposure of Members of the Public

When the commercial nuclear power program in the U.S. started, the basic annual dose limit for nonoccupational exposure was .005 Sv (0.5 rem), and the plants were designed to meet this limit. The actual practice was to provide a margin of safety by designing the plants to keep the instantaneous dose rate from effluents below .005 Sv (0.5 rem) per year. Normal variation in wind direction and other parameters ensured that the actual annual dose would be far below the limit. In addition, 10 CFR Part 20 limited the average dose to the population to .0017 Sv (0.17 rem). This limit did not affect reactor effluents because the doses decline rapidly with increasing distance from the plant. A population dose limit was envisioned for other activities such as "Plowshare," the peaceful uses of nuclear explosives. The Atomic Energy Commission withdrew the population dose limit, and required that releases from nuclear power plants be ALARA. The nonspecific ALARA requirement was a challenge to reactor manufacturers and architect engineers. Consequently, the Atomic Energy Commission undertook the development of criteria that would satisfy the ALARA requirement. The result was the current regulations in 10 CFR 50.34a and Appendix I to 10 CFR Part 50. Acceptable numerical ALARA criteria in Appendix I are:

- From liquid effluents, the annual dose commitments shall be no more than .00003 Sv (0.003 rem) to the whole body or .0001 Sv (0.01 rem) to any organ.

- From airborne effluents, the annual dose commitments shall be no more than 0.01 cGy (0.01 rad) to air from gamma radiation or 0.02 cGy (0.02 rad) to air from beta radiation, and .00005 Sv (0.005 rem) to the whole body or .00015 Sv (0.015 rem) to any organ

- In addition to meeting these criteria, reduce the population dose as much as feasible at a cost not to exceed a figure of merit of $10 per person-Sv ($1000 per person-rem) avoided. (Current guidance suggests $2000 per person-rem.)

Appendix I also specified effluent monitoring, environmental monitoring, investigations, land-use censuses, and reporting.

Appendix I was issued in 1975, but was not fully implemented for all operating reactors for several years. Data from programs that resulted from this appendix were documented for both releases (NUREG/CR-2907, Vol. 17, 1995) and dose (NUREG/CR-2850, Vol. 14, 1996) for 17 years, from 1975 through 1992. Over this period, the energy generation by U.S. nuclear power plants *increased* by a factor of 3.6, while the total annual dose from releases from nuclear power plants to the U.S. population *decreased* by a factor of 27.6 (NUREG/CR-2850, Vol 14, 1996).

ARTICLE 16. EMERGENCY PREPAREDNESS

1. Each Contracting Party shall take the appropriate steps to ensure that there are onsite and offsite emergency plans that are routinely tested for nuclear installations, and cover the activities to be carried out in the event of an emergency.

    For any new nuclear installation, such plans shall be prepared and tested before [the installation] commences operation above a low power level agreed [to] by the regulatory body.

2. Each Contracting Party shall take appropriate steps to ensure that, insofar as they are likely to be affected by a radiological emergency, its own population and the competent authorities of the States in the vicinity of the nuclear installation are provided with appropriate information for emergency planning and response.

3. Contracting Parties that do not have a nuclear installation on their territory, insofar as they are likely to be affected in the event of a radiological emergency at a nuclear installation in the vicinity, shall take appropriate steps for the preparation and testing of emergency plans for their territory that cover the activities to be carried out in the event of such an emergency.

This section discusses (1) emergency planning and emergency planning zones; (2) offsite emergency planning and preparedness; (3) emergency classification system and action levels; (4) responsibilities of the plant operator, state, and local governments; (5) emergency response centers; (6) recommendations for protection in severe accidents; (7) inspection practices and regulatory oversight; (8) the Federal response; and (9) international arrangements.

16.1 Background

The NRC's responsibilities for radiological emergency preparedness are derived from NRC licensing functions under the Atomic Energy Act of 1954 and the Energy Reorganization Act of 1974. Both statutes confer broad regulatory powers on the Commission, and specifically authorize the agency to promulgate regulations that it deems necessary to fulfill its responsibilities under the Acts. Following the accident at Three Mile Island Unit 2 (TMI-2) in March 1979, the regulations were amended to require significant changes in emergency planning and preparedness for commercial nuclear power plants in the United States. The NRC's emergency planning regulations are now considered to be an important part of the regulatory framework for protecting public health and safety, and have been adopted as an added conservatism to the NRC's defense-in-depth safety philosophy of multiple-barrier containment and redundant safety systems. Before a full-power operating license can be issued, NRC regulations require a finding that there is reasonable assurance that adequate measures to protect public health and safety can and will be taken in a radiological emergency (10 CFR 50.47(a)).

16.2 Basis for Emergency Planning and Emergency Planning Zones

Emergency planning in the U.S. recognizes that there is a spectrum of accidents that could exceed the design-basis accidents that nuclear plants are required to accommodate without significant public health and safety impacts. For design-basis accidents, the small releases that might occur would not likely require responses such as evacuating or sheltering the general public. These actions only become important in considering accidents that are much less probable than design-basis accidents. NUREG-0396, "Planning Basis for the Development of State and Local Government Radiological Emergency Response Plans in Support of Light-Water Nuclear Power Plants," and NUREG-0654, "Criteria for Preparation and Evaluation of Radiological Emergency Response Plans and Preparedness in Support of Nuclear Power Plants," describe the emergency planning basis. NUREG-0396 determined that emergency plans should provide for the following outcomes:

- substantial reduction in early severe health effects (injury or death) in a worst-case core melt accident

- dose savings for a spectrum of accidents

The overall objective of emergency response plans is to provide dose savings (and, in some cases, immediate life saving) for a spectrum of accidents that could produce offsite doses in excess of Protective Action Guides. (A Protective Action Guide is the projected dose from an unplanned release of radioactive material at which a specific protective action to reduce or avoid that dose is recommended.)

Following the accident at Three Mile Island Unit 2 (TMI-2), the NRC amended its regulations to require the establishment of two emergency planning zones around a nuclear power plant (10 CFR Part 50, Appendix E.I). Emergency planning zones are defined as the areas for which planning is needed to ensure that prompt and effective actions can be taken to protect the public in the event of an accident. The choice of the size of each emergency planning zone is a judgment regarding the extent of detailed planning that must be performed to ensure an adequate response. In a particular emergency, protective actions might be restricted to a small part of the planning zones. On the other hand, for the worst possible accidents, protective actions might need to be taken beyond the planning zones.

The first emergency planning zone, called the "plume exposure pathway," is an area about 10 miles (16 km) in radius from the center of the plant. The major protective actions planned for this zone, evacuation and sheltering, would be taken to reduce fatalities and injuries from exposure to the radioactive plume from the most severe core melt accidents, and to limit unnecessary radiation exposures to the public from less severe accidents. The second zone, called the "ingestion pathway," is an area about 50 miles (80 km) in radius from the center of the plant. The major protective actions planned for this zone, putting livestock on stored feed and controlling food and water, would be taken to reduce exposure to the public from ingestion of contaminated foodstuffs.

The detailed planning required for the 10 and 50-mile emergency planning zones provides a substantial base for expanding response efforts when warranted by the conditions of a particular accident. Although the plume exposure pathway zone is generally circular, the actual shape is determined on the basis of local factors, such as demography, topography, access routes, and governmental jurisdictional boundaries at a particular site. Smaller emergency planning zones have been established in the U.S. for gas-cooled power reactors and water-cooled power reactors of less than 250 MW.

16.3 Offsite Emergency Planning and Preparedness

The accident at TMI-2 revealed that much better coordination and more comprehensive emergency plans and procedures were needed if the NRC and the public were to have confidence in the readiness of onsite and offsite emergency response organizations to respond to a nuclear emergency. Participation by State and local governments in emergency planning for nuclear power plants in the U.S. was, and still remains, largely voluntary. Before the accident, there had been no clear obligation for the State and local governments to develop emergency plans for radiological accidents, and the Federal role was one of assistance and guidance. After the accident, the NRC amended its emergency planning regulations to require, as a condition of licensing, that each applicant and licensee submit the radiological emergency response plans of State and local governments that are within the plume exposure zone, as well as the plans of State governments within the ingestion pathway zone (10 CFR 50.33(g) and 50.54(s)).

In December 1979, the President directed the Federal Emergency Management Agency (FEMA) to take the lead in ensuring the development of acceptable State and local offsite emergency plans and activities for nuclear power facilities. FEMA's role and responsibilities were subsequently codified in NRC and FEMA regulations and in a Memorandum of Understanding between the two agencies.

FEMA provides its findings regarding the acceptability of the offsite emergency plans to the NRC, the agency that has the ultimate responsibility to determine the overall acceptability of radiological emergency plans and preparedness for a nuclear power reactor. The NRC will not issue a license to operate a nuclear power reactor unless it finds that the state of onsite and offsite emergency preparedness provides reasonable assurance that adequate protective measures can and will be taken in a radiological emergency. The NRC bases its finding on a review of the FEMA findings and determinations as to whether State and local emergency plans are adequate and capable of being carried out, and on its own assessment of whether the onsite emergency plans are adequate and capable of being implemented (10 CFR 50.47(a)).

The principal guidance for preparing and evaluating radiological emergency plans for licensee and State and local government emergency planners is NUREG-0654/FEMA-REP-1, Rev. 1, a joint NRC and FEMA document. NUREG-0654 gives evaluation criteria for meeting the emergency planning standards in the NRC's and FEMA's regulations (10 CFR 50.47(b) and 44 CFR Part 350, respectively). These criteria provide a basis for licensees and State and local governments to develop acceptable emergency plans.

The onsite emergency plan covers all hazards that could potentially affect the safe operation of the plant. Thus, such initiating events as fires, toxic chemical releases, aircraft crashes, and severe natural events are included in a licensee's emergency action level schemes, which are used to classify emergencies and initiate notification of and response to the event. The offsite plans specifically focus on radiological emergency preparedness for the emergency planning zones, although the plans can be, and have been, used for non-radiological incidents that occur in offsite jurisdictions.

The NRC and FEMA coordinate their efforts in evaluating periodic emergency response exercises, which are required by 10 CFR Part 50, Appendix E.F.2, to be conducted every 2 years at all operating nuclear power plant sites. These full-participation exercises are integrated efforts by the licensee and State and local radiological emergency response organizations that have a role under the plan. The NRC evaluates the licensee's performance, and FEMA evaluates the response by State and local agencies. In some cases, various Federal response agencies also participate in these exercises. Any weaknesses or deficiencies that are identified by the NRC or FEMA as a result of the exercise must be corrected through appropriate remedial actions. Besides the biennial exercise of the plume exposure pathway plans, States are required to participate in an ingestion pathway exercise every 6 years with a nuclear power plant located within the States. There is no requirement to involve members of the public in any of the emergency preparedness exercises.

16.4 Emergency Classification System and Emergency Action Levels

NRC regulations establish four classes of emergencies in order of increasing severity. Specifically, these are (1) Unusual Event, (2) Alert, (3) Site Area Emergency, and (4) General Emergency. The specific class of emergency is declared on the basis of plant conditions that trigger the emergency action levels, as discussed below. Typically, licensees have established specific procedures for carrying out emergency plans for each class of emergency. These procedures are to be implemented by the control room staff upon the declaration of an emergency. The event classification initiates all appropriate actions for that class, including notification of offsite authorities, activation of onsite and offsite emergency response organizations, and, where appropriate, protective action recommendations for the public. These same emergency classes are also found in the State and local plans that support each nuclear power plant.

NUREG-0654 gives examples of initiating conditions for each of the four emergency classes. These conditions form the basis for each licensee to establish specific indicators, called emergency action levels. Exceeding these levels indicates that a given initiating condition has been met, and that the proper class of emergency must be declared. An emergency action level is a predetermined, site-specific, observable threshold for an initiating condition that places the plant in a given emergency class. An emergency action level can be an instrument reading; an equipment status indicator; a measurable parameter (onsite or offsite); a discrete, observable event; results of analyses; entry into specific emergency operating procedures; or other factors, including the judgment of plant operators, which, if they occur, indicate entry into a particular emergency class. These levels provide a clear basis for rapidly identifying a possible problem and notifying the onsite emergency response organization and the offsite authorities that an emergency exists. Under NRC regulations, the licensee and State and local governmental authorities must discuss and agree upon the levels, and the NRC must approve them. In Regulatory Guide 1.101, "Emergency Planning and Preparedness for Nuclear Power Reactors" Rev. 3, August 1992, the NRC endorsed the guidance in NUMARC/NESP-007, "Emergency

Planning and Preparedness for Nuclear Power Reactors," Rev. 2, January 1992, as an acceptable alternative to develop emergency action levels.

16.5 Responsibilities of Nuclear Power Plant Operator

In an emergency, the nuclear power plant operator (the licensee) has the primary functions to: (1) control the event and (2) notify offsite officials and, where appropriate, (3) recommend protective actions for the public off site. The plant operator's first priority is to protect the core by ensuring that crucial safety functions are maintained by (1) making and keeping the core subcritical, (2) keeping the water flowing through the core, (3) keeping the core covered with water, (4) providing makeup for water that is boiled off, and (5) removing decay heat from the core to an outside heat sink. The plant operator must also act to prevent or reduce offsite consequences by (1) maintaining reactor containment and the engineered safety features systems, (2) controlling radionuclide releases, and (3) recommending appropriate protective actions to offsite officials.

In parallel with attempts to correct the problem, the plant operator must notify offsite officials of an emergency declaration within 15 minutes (10 CFR Part 50, Appendix E, Section IV.D) and the NRC immediately thereafter. On the basis of the operator's understanding of the reactor core and containment conditions, the operator is also required to recommend initial protective actions to offsite authorities for severe reactor accidents with the potential for offsite consequences. If an actual offsite radionuclide release occurs, the operator is responsible for monitoring the release with offsite organizations to ensure that actions recommended to offsite authorities are appropriate (i.e., that initial protective action recommendations and decisions continue to be valid in light of current, actually monitored data).

16.6 Responsibilities of State and Local Governments

State and local governmental authorities are charged with protecting the public from the offsite consequences that might result from a nuclear power plant accident. These organizations are responsible for making decisions about protective actions, and for notifying the public to take protective actions during a severe accident. State and local officials base their decisions on the recommendations of the plant operator and their own assessment of the situation. The plant operator does not have the authority to declare a state of emergency or order an evacuation of the area surrounding the plant; the operator can only recommend protective actions to the proper offsite officials. Those officials must decide whether to notify the public to implement any protective actions.

Once a decision is made to take protective actions, NRC regulations require that there be both administrative and physical means to alert and promptly inform the public in the plume exposure emergency planning zone. The alert and notification systems in that zone typically consist of a siren warning system to alert the public to listen for emergency information offsite. Offsite authorities control the activation of the system. The systems also send out pre-scripted emergency alert messages to inform the public about actions to be taken. The offsite alert and notification systems must be capable of essentially complete initial notification of the public in the plume exposure zone within about 15 minutes after the operator notifies officials that an urgent situation exists that requires prompt protective actions.

16.7 Emergency Response Centers

The onsite responsibility and authority to act during an emergency resides with the shift supervisor in the plant control room until the technical support center or the emergency operations facility is activated. This includes the authority to declare emergencies, notify offsite officials within 15 minutes of event declaration, and recommend protective actions. The NRC must be notified after the appropriate State and local officials are notified, and no later than 1 hour after the emergency is declared. Upon declaration of an emergency, an onsite emergency director is designated to be in charge of the plant's response. This initially is the shift supervisor, the senior person in the control room. Once the staff, properly augmented, is present after declaration of an emergency, this responsibility (and title) normally transfers to the onsite technical support center and then to the emergency operations facility.

The staff of the onsite technical support center for emergency response has access to the plant's technical information, and is responsible for engineering support of reactor operations during an accident. The center also relieves the reactor operators of peripheral duties and communications that do not directly relate to plant operation. Until the emergency operations facility is activated, the center also performs the functions of that facility. The center is usually located close to the control room inside a protected and shielded area to allow fast access for face-to-face discussions with the control room staff.

The emergency operations facility is off site, and is controlled and staffed by the licensee. This facility manages the overall emergency response of the licensee, coordinates radiological and environmental assessment, recommends public protective actions, and coordinates emergency response activities with Federal, State, and local agencies. NUREG-0737, Supplement 1, "Clarification of TMI Action Plan Requirements," dated January 1983, recommends that the facility be 10 to 20 miles (16 to 32 km) from the site. However, some may be closer (with proper shielding and a backup), and others may be farther from the site. The facility provides for accommodating Federal, State, and local responders to an emergency.

State and local response organizations establish one (or more) emergency operation center(s) upon activation. Some sites have several local governments within the plume exposure zone and, each might have a center. State governments typically operate from emergency operation centers located in the State capital or in a regional office. In some cases, the State may establish a forward emergency operation center near the plant site. The specific location of the off site officials for making emergency management decisions varies, and is very site specific.

16.8 Recommendations for Protective Action in Severe Accidents

The technical basis and guidance for determining protective actions in the U.S. for severe (core damage) reactor accidents are given in NUREG-0654, "Criteria for Protective Action Recommendations for Severe Accidents," Supplement 3, and EPA 400-R-92-001, "Manual of Protective Action Guides and Protective Actions for Nuclear Incidents." These documents reflect the conclusions that have been developed from severe accident studies, such as NUREG-1150, "Severe Accident Risks: An Assessment for Five U.S. Nuclear Power Plants."

Only a very severe reactor accident that causes core damage and failure of the containment could result in doses that are sufficient to cause early significant health effects (injuries or deaths). For a severe accident, the control room staff should know that (1) the core is damaged and (2) a large amount of fission products in the containment atmosphere could be released if the containment fails. However, under severe accident conditions, the performance of the containment may not be predicted with great certainty.

Studies of severe reactor accidents and the effectiveness of protective actions support the following conclusions:

• To substantially reduce risk, evacuation must begin before, or shortly after, a release.

• Movement of even short distances substantially reduces risk.

• Sheltering close to the plant for long periods may not be an effective protective action.

The initial, early protective actions to be recommended to the public under a given set of emergency conditions are predetermined during planning. However, adjustments to preplanned actions may be warranted by specific local conditions. Early evacuation of nearby areas is the most beneficial protective action and, for the most severe accidents, early evacuation is the only available protective action to achieve the basic objectives for radiation protection near the plant (i.e., within about 2 to 3 miles (3.2 to 4.8 km)).

The NRC has incorporated this information into its guidance for response procedures and training manuals for the NRC staff (NUREG/BR-0150, "Response Technical Manual 96.") In addition, the NRC (with the assistance of FEMA) has issued guidance for licensees and offsite authorities regarding protective actions for severe accidents in a draft supplement to NUREG-0654 (NUREG-0654, Supp. 3, "Criteria for Protective Action Recommendations for Severe Accidents," July 1996). The NRC's guidance on evacuation and sheltering in the event of a nuclear power plant accident is consistent with guidance in IAEA TECDOC-953, "Method for the Development of Emergency Response Preparedness for Nuclear or Radiological Accidents," and TECDOC-955, "Generic Assessment Procedures for Determining Protective Actions During a Reactor Accident."

The NRC considers evacuation and sheltering to be the two primary protective actions and prefers prompt evacuation for the population near a plant in a severe reactor accident. However, a supplemental protective action for the general population involves using the thyroid-blocking agent potassium iodide.

After years of deliberation, the NRC has recently amended its regulations for emergency planning. This amendment, "Consideration of Potassium Iodide in Emergency Plans," dated January 2001, requires that each State consider giving potassium iodide as a protective measure to the general public, as a supplement to evacuation and sheltering. The Commission found that potassium iodide is a reasonable, prudent and inexpensive supplement to evacuation and sheltering for specific local conditions. In addition, the NRC has agreed to fund a supply of potassium iodide for States or, in some cases, local governments, that choose to give potassium iodide to the general public as part of their emergency plans. The NRC is also working with FEMA and the other member agencies to update the 1985 Federal Radiological Emergency Response Plan policy on potassium iodide.

In parallel with the new NRC rule, the U.S. Food and Drug Administration (FDA) is updating its guidance on potassium iodide prophylaxis. FDA issued proposed guidance in January 2001 which would set 5 centigray as the level at which potassium iodide prophylaxis should be carried out. Final FDA guidance is expected by the end of 2001. The NRC is also working with FEMA to update the previous federal policy statement on potassium iodide prophylaxis to reflect the new NRC rule and the updated FDA guidance. Finally, the Commission is working with EPA on an update of EPA's protective action guidelines, which will among other things update the potassium iodide prophylaxis protective action guide to be consistent with final FDA guidance.

16.9   Inspection Practices — Regulatory Oversight Process for Emergency Preparedness

The NRC's Revised Reactor Oversight Process, described in Article 6, provides for emergency preparedness. Specifically, the process allows the licensee latitude in managing emergency preparedness programs, including corrective actions, as long as the performance indicators and inspection findings are within an acceptable performance band.

Emergency preparedness is the final barrier between reactor operations and protection of public health and safety. As such, emergency preparedness is a major component of the regulatory oversight process, and is one of the seven recognized cornerstones of safety in the process. A basic premise of the regulatory framework is that there is a threshold of licensee safety performance above which the NRC can allow licensees to address weaknesses with NRC oversight through a risk-informed inspection process. The regulatory process for the cornerstone for emergency preparedness established the cornerstone objective to: "Ensure that the licensee is capable of implementing adequate measures to protect the public health and safety during a radiological emergency." Oversight of this emergency preparedness cornerstone is achieved through three performance indicators and a supporting risk-informed inspection program. The most risk-significant areas of emergency preparedness have been identified as classification, notification, protective action recommendation, and assessment.

The emergency preparedness performance indicators are as follows:

•   Drill and Exercise Performance

•   Emergency Response Organization Drill Participation

•   Alert and Notification System Reliability

The performance indicator for Drill and Exercise Performance monitors timely and accurate licensee performance in drills, exercises, and actual events when presented with opportunities to classify emergencies, notify offsite authorities, and recommend protective actions. The indicator for Emergency Response Organization Drill Participation measures the percentage of key members of the licensee's emergency response organization who have participated in proficiency enhancing drills, exercises, training opportunities, or an actual event over a certain

16-8

time. The Alert and Notification System Reliability indicator monitors the reliability of the offsite alert and notification system, which is a critical link for alerting and notifying the public of the need to take protective actions.

The inspectable areas of the cornerstone for emergency preparedness under the Reactor Oversight Process include the following:

- Problem Identification and Resolution: Inspectors evaluate the licensees' programs for problem identification and resolution as they relate to the emergency preparedness program.

- Drill and Training Evolution Observation: Inspectors evaluate drills and simulator-based training evolutions in which shift operating crews participate.

- Biennial Exercise: The Inspectors independently observe the licensee's performance in classifying, notifying, and developing recommendations for protective actions, and other activities during the exercise. The inspectors also ensure that the licensee's critique is consistent with their observations.

- Alert and Notification System: One of the three performance indicators addresses performance of the alert and notification system. However, for the statistics to be valid, the testing program must be conducted in accordance with program procedures. Consequently, the inspectors verify the compliance of the testing program with program procedures.

- Emergency Action Level Revision Review: Inspectors review all of the licensee's changes to emergency action levels to determine if any of the changes have decreased the effectiveness of the emergency plan.

- Emergency Response Organization Augmentation: Inspectors review the augmentation system to determine whether, as designed, it will support augmentation of the emergency response organization in accordance with the goals for activating the emergency response facility.

- Performance Indicator Verification: Inspectors verify that the data reported for the performance indicator values are valid.

- Emergency Plan Changes: Inspectors sample changes to the emergency plan to ensure that the effectiveness of the emergency plan has not decreased.

As explained in Article 6, the NRC has a formal way of dispositioning inspection findings, called the significance determination process. Its intent is to normalize inspection findings to a level of risk that is similar to the thresholds for performance indicators. An inspection finding may be in the licensee response band, in which case, it is placed in the licensee's problem identification and resolution program. However, the NRC follows up more risk-significant findings with actions that are appropriate for the finding. The NRC has developed a predictable way of escalating regulatory engagement to ensure that agency participation in resolving licensee problems is risk-informed. Specifically, the NRC documents violations of regulatory requirements, but normally leaves items that are not risk significant to the licensee's problem identification and resolution program. The agency only levies civil penalties for events that have actual consequences. Under the emergency preparedness cornerstone, such events include a failure to identify, notify, and/or communicate protective action recommendations during a site area emergency or general emergency.

Although FEMA has no direct regulatory authority over State and local governments, and the evaluators of FEMA exercises are not considered inspectors, FEMA's exercise findings carry great weight in the NRC's regulatory process. State governments and the NRC are notified of significant deficiencies in offsite performance shortly after the exercise, and FEMA issues a formal exercise report about 90 days after the exercise. This report identifies FEMA's exercise findings, and the findings are expected to be closed either before or during the next exercise. Because of the potential effect of deficiencies on offsite emergency preparedness, they are expected to be corrected within 120 days of the exercise. Failure of off site organizations to correct deficiencies in a timely manner could lead to FEMA's withdrawal of its finding of "reasonable assurance."

16.10 Federal Response to an Emergency

The Federal Government's ability to respond to a radiological emergency has significantly improved since the accident at TMI-2. Improvements include the establishment of a Federal response plan, direct communication links between nuclear power plants and the NRC, a designated staff of full-time operations officers at the NRC, and a direct near-real-time electronic data link between the licensee's onsite computer system and the NRC that automatically transmits selected plant parameters.

The NRC recognizes the nuclear power plant operator (licensee) and the State or local government as the two primary decisionmakers in a radiological emergency at a licensed power reactor. The operator is primarily responsible for mitigating the consequences of an incident, and recommending timely and proper protective actions to State and local authorities. The State or local governments are ultimately responsible for implementing proper protective actions for public health and safety.

The Federal response is primarily designed to support the efforts of the facility operator and offsite officials. For an emergency with potential radiological consequences, Federal response activities are conducted in accordance with the "Federal Radiological Emergency Response Plan," which is promulgated in Section V of NUREG/BR-0230, "Response Coordination Manual 96." This document describes the role of the lead Federal agency and other Federal agencies that may respond to a radiological emergency. The NRC is the lead Federal agency for a radiological emergency that applies to NRC-licensed nuclear facilities or materials. The lead Federal agency's principal roles in an emergency at a nuclear power plant are to (1) ensure that the licensee has properly communicated a recommendation for protective actions to State and local authorities, (2) independently assess reactor conditions, the progression of the accident, and possible consequences, and inform the State authorities of these results, and (3) coordinate the Federal response to the emergency to support the State government.

In fulfilling its legislated mandate for protecting the public health and safety, the NRC has developed a plan and procedures that detail the NRC's response to incidents involving licensed material and activities (NUREG-0728, "NRC Incident Response Plan"). In accordance with that plan, the NRC will initially assess any reported accident, and decide if it will respond as an agency and dispatch an NRC Site Team from its regional offices. The NRC will dispatch a team to the site for all serious accidents.

Once the NRC has decided to respond as an agency, the agency's Operations Center is activated in the Washington DC area, and the Regional Incident Response Center is activated. The NRC Operations Center will then (1) maintain continuous communications with the facility, (2) assess the accident, (3) advise the facility operator and offsite officials, (4) coordinate the Federal radiological response with other Federal agencies (FEMA, DOE, EPA, etc.), and (5) respond to inquires from the national media from a media center. The NRC Operations Center will direct the NRC response for about 4–8 hours until the lead is transferred to the NRC Site Team. The NRC Operations Center will be staffed with experts on the facility and its emergency plans, response to accidents, and media and governmental relations.

As soon as the NRC Site Team arrives at the scene and declares that it is ready to assume the agency's leadership role, it is given the authority to direct the Federal response. The NRC Site Team then sends representatives to centers that are used by the facility and offsite officials to coordinate the response. The NRC Site Team has access to extensive radiological monitoring capabilities through the U.S. Department of Energy, including field teams and airborne monitoring. The Federal radiological monitoring efforts to support State and local officials are coordinated at the Federal Radiological Monitoring and Assessment Center. The NRC Site Team also sends representatives to the Joint Information Center established by the facility or local government, where all media inquires are addressed.

The NRC, in cooperation with other Federal agencies, has developed and published tools and computer codes for assessing radiological accidents (NUREG/BR-0150 and RASCAL 3.0.1) and coordinating the agency's response (NUREG/BR-0230). The technical tools are designed for use under IAEA TECDOC-955. The NRC also regularly participates in exercises of its response program, participating fully in about four nuclear power plant exercises and one fuel cycle facility exercise each year and to a lesser extent in numerous other exercises.

The Federal Government's ability to respond to a terrorist threat or act at a domestic nuclear facility has significantly improved since the bombing of the Federal building in Oklahoma City in April 1995. Issued in June 1995, Presidential Decision Directive 39, designates the Federal Bureau of Investigation (FBI), through the Department of Justice, and FEMA as the lead federal agencies for responding to domestic terrorism. Terrorist threats or acts at a nuclear facility are assumed to be radiological emergencies as well. In responding to the radiological emergency, the NRC continues in its role as the lead Federal agency while supporting both FBI and FEMA in their counter terrorism roles. The NRC participated in two counter terrorism exercises in 2000 and one in 2001 involving NRC licensed facilities. The NRC is revising its Incident Response Plan, NUREG-0728, in several areas, and this revision will further clarify and enhance the Federal response to terrorism at NRC licensed facilities.

16.11 International Arrangements

The NRC has agreements with Canada and Mexico and commitments to the IAEA.

The NRC has signed agreements with Canada and Mexico under which it will promptly notify and exchange information in the event of an emergency that has the potential for trans-boundary effects. The agreement with Canada is the "Agreement Between the Government of the United States of America and the Government of Canada on Cooperation in Comprehensive Civil Emergency Planning and Management." It is implemented by the procedure specified in

"Administrative Arrangement Between the United States Nuclear Regulatory Commission and the Atomic Energy Control Board of Canada for Cooperation and the Exchange of Information in Nuclear Regulatory Matters." (Both documents are dated June 21, 1989.)

The agreement with Mexico is the "Agreement for the Exchange of Information and Cooperation in Nuclear Safety Matters." It is implemented by the "Implementing Procedure for the Exchange of Technical Information and Cooperation in Nuclear Safety Matters Between The Nuclear Regulatory Commission of the United States of America and the Commission National de Seguridad Nuclear y Salvaguardias of Mexico." (Both documents are dated October 6, 1989.)

To meet the U.S. commitment under the IAEA "Convention on Early Notification of a Nuclear Accident," the NRC will promptly notify the IAEA if a serious accident occurs at a commercial nuclear power plant. After this initial notification, the NRC will work with the U.S. Department of State to update the IAEA.

In May 2001, the Commission approved the full participation of the U.S. in the international nuclear event scale. Full participation is defined as the evaluation of all nuclear events that fall under the regulatory purview of the Agreement States by the NRC staff for possible rating on the international nuclear event scale. Full participation will allow international stakeholders to quickly grasp the significance of U.S. events.

The international nuclear event scale is designed to provide a consistent means to convey the significance of a wide range of reactor and materials events to the international community. The NRC has participated in a very limited manner since December 1992. Since that time, the scope and level of worldwide participation in the program has expanded, primarily as the result of efforts by the IAEA and the Nuclear Energy Agency of the Organization for Economic Cooperation and Development. The NRC has arranged for training of appropriate NRC staff members by the IAEA in early November 2001 and expects that the U.S. will be fully participating by December 31, 2001.

ARTICLE 17. SITING

Each Contracting Party shall take the appropriate steps to ensure that appropriate procedures are established and implemented for

(i) evaluating all relevant site-related factors that are likely to affect the safety of a nuclear installation for its projected lifetime

(ii) evaluating the likely safety impact of a proposed nuclear installation on individuals, society, and the environment

(iii) re-evaluating, as necessary, all relevant factors referred to in sub-paragraphs (i) and (ii) so as to ensure the continued safety acceptability of the nuclear installation

(iv) consulting Contracting Parties in the vicinity of a proposed nuclear installation, insofar as they are likely to be affected by that installation and, upon request, providing the necessary information to such Contracting Parties, in order to enable them to evaluate and make their own assessment of the likely safety impact on their own territory of the nuclear installation

This section explains NRC's responsibilities for siting: site safety, environmental protection, and emergency preparedness. First, this section discusses the regulations applying to site safety and their implementation. It emphasizes regulations applying to seismic, geological, and radiological assessments. Next, it explains environmental protection. Emergency preparedness is discussed in Article 16, "Emergency Preparedness." International arrangements, which would apply to Contracting Parties in obligation (iv), above, are also discussed in Article 16.

17.1 Background

The NRC's responsibilities for siting stem from the Atomic Energy Act (AEA) of 1954, the Energy Reorganization Act (ERA) of 1974 (as discussed earlier) and the National Environmental Policy Act (NEPA) of 1969. These statutes confer broad regulatory powers on the Commission, and specifically authorize the NRC to promulgate regulations that it deems necessary to fulfill its responsibilities under the Acts.

The NRC's siting regulations are integral to protecting public health and safety and the environment. Siting away from densely populated centers has been, and will continue to be, an essential component of the NRC's defense-in-depth safety philosophy, which also includes multiple-barrier containment and redundant safety systems. The primary factors that determine public health and safety are the reactor design, and construction and operation of the facility. However, siting factors and criteria are important in ensuring that radiological doses from normal operation and postulated accidents will be acceptably low, natural phenomena and man-made hazards will be properly accounted for in the design of the plant, and the human environment will be protected during the construction and operation of the plant.

17.2 Safety Elements of Siting

This section explains the safety elements of siting. First, it discusses seismic and geological assessments. It then discusses radiological assessments performed for initial licensing, as a result of facility changes, and according to regulatory developments that have occurred since the licensing of all U.S. operating plants.

17.2.1 Background

The NRC's site safety regulations consider societal and demographic factors, man-made hazards (such as airports and dams), and physical characteristics of the site (such as seismic and meteorological factors) that could affect the design of the plant. The requirements are specified in 10 CFR Part 100, "Reactor Site Criteria," issued in 1962, Appendix A, "Seismic and Geologic Siting Criteria for Nuclear Power Plants," to 10 CFR Part 100, added in 1973, and 10 CFR Part 100.23 "Geologic and Seismic Siting Criteria," added in 1997.

The applicant's safety analysis report is required to describe characteristics in and around the site, and contain accident analyses that are relevant to evaluating the suitability of a site. A number of regulatory guides provide guidance regarding issues of site safety that applicants need to address. NUREG-0800, "Standard Review Plan for the Review of Safety Analysis Reports for Nuclear Power Plants," guides the staff in reviewing the site safety content of these reports.

Notably, once licensed to operate, the licensee is expected to monitor the environs around the nuclear power plant, and report changes in the environs in its safety analysis report that may affect the continued safe operation of the facility. Such changes may affect the regulatory status of the facility in different ways. For example, if population characteristics near the plant change during the plant's operating lifetime, the licensee must continue to assess the emergency planning criteria — rather than the siting criteria — to ensure the licensee's continued ability to carry out measures to protect the public. As another example, if the plant experiences an extreme natural event (such as an earthquake or flood), the licensee must demonstrate that no functional damage has occurred to those features that are deemed necessary for continued safe operation without undue risk to the public, and that the licensing basis is maintained.

17.2.2 Assessments Applying to Seismic and Geological Aspects of Siting

Appendix A to 10 CFR Part 100 describes the investigations required to obtain the geologic and seismic data necessary to determine site suitability and to provide reasonable assurance that a nuclear power plant can be constructed and operated at a proposed site. It also describes procedures for determining the design basis for vibratory ground motion from earthquakes at a site. The approach given in this appendix for determining the ground motion for a safe shutdown earthquake is deterministic, according to which an applicant develops a single set of earthquake sources, and for each source, postulates an earthquake as the source of ground motion that can affect the site, locates the postulated earthquake according to prescribed rules, and then calculates ground motion at the site. Used for the past 20 years, this approach has been considered to be suitably conservative, but has the shortcoming of not explicitly recognizing certain uncertainties about earthquake phenomena. Because of these uncertainties, experts have differed about the largest earthquakes to be considered and ground motion models to be used in the licensing process.

Over the past decade, analysts have developed and used probabilistic methods of analysis that incorporate various models for earthquake size, ground motion, and other parameters. These methods have the advantage of not only incorporating various models (and data), but also weighting them on the basis of judgments as to their validity. Thus, these methods provide an explicit expression for the uncertainty in ground motion estimates and a means of assessing sensitivity to various input parameters.

In 1997, the NRC added Section 100.23, "Geologic and Seismic Siting Criteria," to the regulations, defining the principal geological and seismic considerations that guide the Commission in evaluating the suitability of a proposed site and adequacy of the design bases. Explicitly recognizing that there are inherent uncertainties in establishing the seismic and geologic design parameters, Section 100.23 allows licensees to address these uncertainties through a probabilistic seismic hazard analysis or suitable sensitivity analyses.

Regulatory Guide 1.165, "Identification and Characterization of Seismic Sources and Determination of Safe Shutdown Earthquake Ground Motion," describes methods acceptable to the NRC staff for implementing the regulation. Standard Review Plan Section 2.5.2, "Vibratory Ground Motion," Revision 3 guides the staff in its reviews.

17.2.3 Assessments of Radiological Consequences for Initial Licensing

10 CFR Part 100, issued in 1962, is the regulation under which all U.S. operating plants were licensed. It contains provisions for assessing whether radiological doses from postulated accidents will be acceptably low. To evaluate a proposed site, 10 CFR Part 100 stated that an applicant should assume a fission product release from the core, the expected demonstrable leak rate from the containment, and the meteorological conditions for the site to determine the distances for the exclusion area boundary and low population zone.

10 CFR Part 100.3, "Definitions," defines the exclusion area as the area surrounding the reactor over which the licensee has the authority to determine all activities, including exclusion or removal of personnel and property from the area. Residences, although not prohibited, are not expected to be located within the exclusion area boundary. Typically, these boundaries are on the order of hundreds of meters. Every site must have a designated low population zone immediately surrounding the exclusion area boundary so that there is a reasonable probability that appropriate protective measures could be taken in a serious accident. The low population zone must be less than three-fourths of the distance to the nearest densely populated center of 25,000 or more residents. Typically, low population zones are on the order of thousands of meters.

Assumptions about the release of fission product were inextricably linked to information in TID-14844, "Calculation of Distance Factors for Power and Test Reactor Sites," which was published concurrently with the 1962 version of Part 100. Although not part of 10 CFR Part 100, TID-14844 was included as a "Note" to 10 CFR Part 100 as a source of additional guidance to users.

10 CFR Part 100 also stated that the fission product release assumed for the calculations to evaluate a proposed site should be predicated upon a *major accident* that would result in potential hazards that are not exceeded by those from any accident that is considered credible. For some time, such a major accident was termed the "maximum credible accident" and, later, the "design-basis accident." Such accidents have generally been assumed to result in a substantial meltdown of the core, with subsequent release of appreciable quantities of fission products. The assessment of the dose from a design-basis accident applied bounding conditions and simplifying assumptions.

The dose criteria, established in the 1962 version of 10 CFR Part 100, are a total radiation dose to the whole body that would not exceed .25 Sv (25 rem), and a total radiation dose to the thyroid from iodine exposure that would not exceed 3 Sv (300 rem). (These criteria would be revised in a revision to 10 CFR Part 100, as discussed below.) Each of these doses applies to both a hypothetical individual at the boundary of the exclusion area (for the 2 hours immediately after the onset of a postulated release of fission product), and a hypothetical individual at the outer boundary of the low population zone (over the course of the accident, which is taken to be 30 days). Of these four criteria, with rare exceptions, the limiting criterion for most U.S. nuclear power plants has been the thyroid dose to an individual at the exclusion area boundary. The whole-body dose, as described in the rule, corresponds numerically to the "once in a lifetime" accidental or emergency dose for radiation workers. Neither the use of this dose nor the thyroid dose was to imply that these doses are acceptable as limits for an emergency dose to the public under accident conditions. Rather, they are reference values that could be used to evaluate sites for potential reactor accidents of exceedingly low probability of occurrence and low risk of exposing the public to radiation. The dose criteria are used only together with conservative, worst-case analysis inputs, assumptions, and methods, such as the assumption of full core melt for a design-basis loss-of-coolant accident.

For design-basis accidents other than the loss-of-coolant accident, such as steam generator tube ruptures, the NRC's Standard Review Plan tabulates acceptance criteria for doses that are a fraction or small fraction of those stated in the 10 CFR Part 100. These reduced acceptance criteria reflect the higher probability and lower consequences of these less severe design-basis accidents.

The NRC issued regulatory guidance for licensees to use to implement the requirements regarding radiological criteria of 10 CFR Part 100. Issued in 1970, Regulatory Guide 1.3, "Assumptions Used for Evaluating the Potential Radiological Consequences of a Loss of Coolant Accident for Boiling Water Reactors," and Regulatory Guide 1.4, "Assumptions Used for Evaluating the Potential Radiological Consequences of a Loss-of-Coolant Accident for Pressurized-Water Reactors," applied the assumptions about the source term in TID-14844, and provided additional insights. ("Source term," as used in this article, refers to the release of fission products from the reactor core into the containment during an accident.) These guides also provided dispersion models for onsite meteorological measurement programs. In 1979, the NRC published Regulatory Guide 1.145, "Atmospheric Dispersion Models for Potential Accident Consequence Assessments at Nuclear Power Plants," as improved guidance for developing relative dilution factors for design-basis accident calculations.

17.2.4 Assessments of Radiological Consequences as a Result of Facility Changes

Although applicants perform dose analyses primarily to support reactor siting, licensees are required to evaluate the potential increase in the consequences of accidents that might result from modifying facility systems, structures, and components. Commitments, including the radiological acceptance criteria, made by the applicant during siting and documented in its final safety analysis report, remain binding until modified. Consequently, a licensee must evaluate the potential consequences of design changes against these radiological criteria to demonstrate that the design changes result in a design that still conforms to the regulations and commitments. For example, if the licensee seeks a change in the containment leak rate or filtration efficiency, it would assess the change in dose that could result from that change before implementing it to ensure continued compliance with the dose criteria. If the licensee finds that the consequences increase more than minimally as outlined in 10 CFR 50.59, "Changes, Tests, and Experiments," (or require a change to the Technical Specifications), as discussed in Article 14, it must obtain prior NRC approval for the proposed modification.

17.2.5 Assessments of Radiological Consequences According to Recent Regulatory Developments

This section explains assessments according to regulatory developments that have occurred after the licensing of all U.S. operating nuclear plants. The section begins by explaining the 1996 revision to 10 CFR Part 100. It then discusses the development an alternative source term which culminated in the issuance of NUREG-1465, "Accident Source Terms for Light-Water Nuclear Power Plants." It then explains how the alternative source term can be applied to operating reactors under a recent rule, 10 CFR 50.67, "Accident Source Term." Finally, it describes experience and gives examples.

17.2.5.1 Governing Documents and Process

Revision to 10 CFR Part 100, "Reactor Site Criteria" (1996)

In 1996, the NRC revised 10 CFR Part 100 to consider the substantial additional body of information on fission product releases that had been developed in the years after TID-14844 was issued and, particularly, as a result of severe-accident research after the accident at Three Mile Island in 1979. The revised rule was intended to be flexible so that the regulatory framework could be accommodating (and not subject to repeated revisions) as new insights continued to unfold. As part of that revision, the NRC added a new subpart (Subpart B, "Evaluation Factors for Stationary Power Reactor Site Applications on or After January 10, 1997") to Part 100 and new paragraphs to 10 CFR 50.34, "Contents of Applications," while retaining the previous requirements that remained applicable to reactors that had previously been licensed.

The 1996 revision replaced the dose criteria of the 1962 version with a single value of 0.25 Sv (25 rem) as the total effective dose equivalent (TEDE). The TEDE is defined as the sum of the deep-dose equivalent from external exposure and the committed effective dose equivalent from internal exposure. The TEDE reference value of .25 Sv (25 rem) was justified by the same logic that was applied in the 1962 version of the rule. The shift to the use of TEDE provided consistency with the 1991 changes to 10 CFR Part 20, "Standards for Protection Against

Radiation," as discussed in Article 15. Moreover, the TEDE is convenient for use with revised source terms, reflecting the myriad of nuclides, besides the noble gases and iodine that were traditionally considered, predicted to be released into the containment. The calculation to determine the exclusion area boundary remained predicated on a 2-hour period, but as a performance measure of the efficacy of the design, the revised regulation specified that the calculation was to use the particular 2-hour period that produced the highest dose value. This calculation has been characterized as the "highest," "worst," and "sliding" 2-hour calculation. The calculation for the low population zone remained predicated on the course-of-the-accident or 30-day dose. Assumptions about dispersion remained the same as in the past.

NUREG-1465, "Accident Source Terms for Light-Water Nuclear Power Plants"

In 1995, the NRC published NUREG-1465, "Accident Source Terms for Light-Water Nuclear Power Plants." This document distilled the body of research, particularly research conducted after the accident at TMI-2, into a practical guide to estimate more realistically the source terms being released into containment, including the timing, nuclide types, quantities, and chemical form, given a representative severe core melt accident. This NUREG provides the technical bases for revised source terms and will form the basis for developing regulatory guidance mainly for future reactors. The revised source term is also referred to as an "alternative source term."

The major differences between the source term of TID-14844 and the revised source term of NUREG-1465 are as follows. The source term of TID-14844 has three categories of radionuclides. By contrast, the revised source term has eight categories, grouped on the basis of similarity in chemical behavior. The source term of TID-14844 is founded on an assumption of an immediate release of the activity; in contrast, the revised source term is categorized into phases, depending on the degree of fuel melting and relocation, vessel integrity, and core-concrete interaction. The source term in TID-14844 is also founded on an assumption of iodine being predominantly elemental; by contrast, the revised source term is founded on an assumption that radioiodine is predominantly cesium iodide (CsI), an aerosol that is more amenable to mitigation mechanisms. In addition, the number and variety of radionuclides considered in NUREG-1465 significantly exceed those considered in TID-14844, as does their affinity to affect specific organs.

10 CFR 50.67, "Accident Source Term"

In considering the applicability of an alternative (or revised) source term to operating reactors, the NRC determined that the analytical approach applying TID-14844 would continue to adequately protect public health and safety, and licensees of already licensed reactors would not be required to reanalyze accidents using the revised source term. The NRC also concluded that some licensees may use an alternative source term in analyses to support operational flexibility and cost-beneficial licensing actions. The NRC acted to provide a regulatory basis for licensees to voluntarily amend their facility design bases to allow use of an alternative source term in design-basis analyses. First, the NRC solicited ideas on how to implement an alternative source term. Second, the NRC initiated a comprehensive assessment of the overall effect of substituting the NUREG-1465 revised source term for the traditionally used TID-14844 source term at three typical facilities. Specifically, the objective of this assessment was to evaluate the issues of applying the revised source term at operating plants in the absence of design changes as well as sensitivity analyses to assess such changes. The NRC subsequently reported the results of this assessment in SECY 98-154, "Results of the Revised (NUREG-1465) Source Term Rebaselining for Operating Reactors," dated June 30, 1998. Third, the NRC accepted requests for license amendments to implement the alternative source term at a small number of pilot plants. The NRC reviewed these pilot projects, and incorporated the resulting insights into regulatory guidance. Fourth, the NRC assessed whether rulemaking would be necessary to allow operating reactors to use an alternative source term. As a result of that assessment, the NRC subsequently added to its regulations 10 CFR 50.67, "Accident Source Term."

10 CFR 50.67, which became effective in January 2000, applies to all holders of operating licenses or renewed licenses whose initial operating license was issued before January 10, 1997, under 10 CFR Part 50, "Domestic Licensing of Production and Utilization Facilities." The rule allows interested licensees to propose (by way of a license amendment) replacing the TID-14844 source term with an alternative source term. The rule also applies the TEDE criteria for the exclusion area boundary and low population zone identical to those provided in 10 CFR 50.34, "Contents of Applications," for future plants. In addition, the rule establishes the value of .05 Sv (5 rem) TEDE as the criterion for control room habitability, in lieu of the value originally given in General Design Criterion 19 in Appendix A to 10 CFR Part 50. As guidance to implement this rule, in July 2000, the NRC issued Regulatory Guide 1.183, "Alternative Radiological Source Terms for Evaluating Design-Basis Accidents at Nuclear Power Reactors." This regulatory guide defines an acceptable source term, and provides analysis inputs, assumptions, methods, and acceptance criteria that are acceptable to the NRC for use with the alternative source term.

Under the rule and its supporting guidance, an applicant may voluntarily pursue a full-scope implementation or a selective implementation. A selective implementation consists of using selected characteristics of the alternative source term for a limited plant modification. For example, a licensee may use the alternative source term to support relaxing certain requirements on containment isolation and filtration operability during refueling. In this case, the licensee may only need to analyze a fuel handling accident to support the proposed changes. After obtaining approval of the alternative source term for this selective implementation, the licensee may use the source term (and dose criteria) in later analyses of fuel handling. If this licensee later opted to use the alternative source term for a loss-of-coolant accident, it would have to request an additional amendment. By contrast, if the licensee pursued a full-scope implementation, the alternative source term and the dose criteria would become a permanent part of that facility's design basis, and the licensee could evaluate subsequent design changes using 10 CFR 50.59 criteria. If the licensee met these criteria, it could implement the desired change without further NRC review.

The NRC does not require that every design-basis radiological calculation be reassessed for either a selective or full-scope implementation. Rather, the NRC decided that the whole body and thyroid dose results determined using the TID-14844 source term would always bound the results for the TEDE and the alternative source term, assuming that all other parameters are held equal. As a result, a licensee need not update previous calculations for the change in the source term. Instead, licensees are only required to update calculations for which analysis inputs or assumptions are no longer valid because of the proposed modification for both selective or full-scope implementations. The NRC does require, at a minimum, that a design-basis loss-of-coolant accident be analyzed for a full-scope implementation. Once the NRC approves selective or full-scope implementation that applies an alternative source term, it expects that future revisions to radiological calculations will reflect the revised design-basis source term and TEDE criteria.

17.2.5.2 Experience and Examples

The NRC has applied the 1996 revision to 10 CFR Part 100, along with the revised source term, in its design certification review for a passive advanced light-water reactor. In that instance, the applicant was able to demonstrate to the NRC's satisfaction that it could use this source term and the TEDE dose criteria. In addition, the applicant was able to defend innovative solutions for certain natural removal mechanisms, without conventional engineered safety features, and still not need an excessively large exclusion area.

The NRC has also approved several applications of the alternative source term, and is currently considering several more. Licensees of boiling-water reactors have used the alternative source term to justify increasing the maximum allowable leakage from the main steam isolation valve, relaxing certain operability requirements for the standby gas treatment system, and supporting reactor power uprates. Licensees of pressurized-water reactors have used the alternative source term to justify removing certain charcoal filtration units, to keep the containment personnel access hatch open during refueling, to support reactor power uprates, and to resolve concerns about control room habitability.

17.3 Environmental Protection Elements of Siting

This section explains the environmental protection elements of siting. It covers the governing documents and site approval process. Since the last operating plants in the U.S. have been licensed, issues have arisen that must be considered in siting reviews. This section explains the effect of these issues on siting reviews.

17.3.1 Governing Documents and Process

The environmental protection elements of siting consist of the plant's demands on the environment (e.g., water use and effects of construction and operation). These elements are addressed in 10 CFR Part 51, "Environmental Protection Regulations for Domestic Licensing and Related Regulatory Functions." 10 CFR Part 51 implements the National Environmental Policy Act of 1969 consistent with the NRC's statutory authority, and reflects the agency's policy to voluntarily account for the regulations of the Council on Environmental Quality, subject to certain conditions. The NRC recognizes its continuing obligation to conduct its domestic licensing and related regulatory functions to be receptive to environmental concerns, and to serve responsibly as an independent regulatory agency to protect the radiological health and safety of the public. Integrating environmental reviews into its routine decisionmaking, the NRC considers environmental protection issues and alternatives before taking any action that may significantly affect the human environment.

The site approval process leading to the construction or operation of a nuclear power plant requires the NRC to prepare an environmental impact statement. As part of that effort, the NRC conducts scoping for the proposed action. In general, scoping is open to anyone who requests an opportunity to participate. Scoping consists of defining the proposed action; determining the scope of the environmental impact statement and identifying significant issues; identifying, and eliminating from detailed study, issues that are peripheral, not significant, or have been covered by prior environmental reviews; identifying other environmental assessments or environmental impact statements that are related to but not part of the scope of the environmental impact statement under consideration; identifying other required environmental reviews and consultations; indicating the relationship of the timing of the preparation of the environmental impact statement and the Commission's planning and decisionmaking schedule; identifying cooperating agencies; and describing how the environmental impact statement will be prepared. After scoping, the NRC summarizes the determinations and conclusions, including the significant identified issues.

The NRC conducts an acceptance review of the applicant's environmental report to determine whether it contains complete information to initiate the detailed Commission review. If the application is not complete, the staff specifically requests additional information before docketing the application. If the application is acceptable, the staff initiates its reviews and analyses, which include a site audit, and may request additional detailed information or clarification. Requests for information add to the record of the staff's evaluation of the application at this review stage.

When the staff has reviewed the environmental report, it writes the draft environmental impact statement following the guidance in the Environmental Standard Review Plan, NUREG-1555. Depending on the licensing application, the draft environmental impact statement may also be a supplement to an existing environmental impact statement or generic environmental impact statement. When writing the draft statement, the NRC addresses comments regarding the applicant's environmental report, which may be received from Federal, State, regional, local, and affected Native American tribal agencies during the scoping and audit activities.

The staff issues the draft environmental impact statement to the public, mainly as a summary of the staff's initial conclusions about an application. The draft discusses the proposed action, and the staff's assessment of its potential benefits and environmental costs. This discussion is intended to give the public an opportunity to comment, request clarification, recommend changes, or provide additional information for the staff to consider in assessing the cost-benefit balance. If the staff does not receive any comments, it can publish the draft statement as a final environmental impact statement. The staff must consider and disposition comments before issuing the draft as a final environmental impact statement. The final environmental impact statement is a summary of the evaluation of the environmental part of the application on the anticipated impact of the proposed action on the environment. It is provided to the public and is used as the main body of environmental evidence at any public hearing, should one be held, to support the Commission's conclusion that the proposed action should be approved or rejected.

Since nuclear power plants that are located near territorial boundaries may affect the physical environment of adjacent countries, the draft environmental impact statement is given to governmental organizations for comment in those countries. If the staff receives comments, it responds to them in its final environmental impact statement. Responses to comments may take one or more forms, including no change, a change in a part of the draft environmental impact statement, or the addition of new material to the relevant section identified in the discussion of comments.

The updated and revised environmental standard review plans (NUREG-1555) guide the staff's environmental reviews for a range of applications, including "green field" (i.e., undeveloped sites) reviews for construction permits and operating licenses in 10 CFR Part 50, for early site permits in 10 CFR Part 52, Subpart A, and for combined licenses in 10 CFR Part 52, Subpart C, when the application does not reference an early site permit. These governing documents and processes are discussed in detail in Article 19. Environmental standard review plans are also appropriate for use in environmental reviews of applications for combined licenses in 10 CFR Part 52, Subpart C, when the applications reference an early site permit. Reviews of early site permit applications are limited in the sense that (1) the reviews focus on the environmental effects of reactor construction and operation that have characteristics that fall within the postulated site parameters, and (2) the reviews need not assess benefits (for example, the need for power). The environmental reviews of applications for combined licenses that reference an early site permit are limited to consideration of (1) information to demonstrate that the design of the facility falls within the parameters specified in the early site permit, and (2) any significant environmental issue that was not considered in any previous proceeding on the site or design.

The environmental standard review plans in NUREG-1555, Supplement 1 guide the staff's environmental review for license renewal applications under 10 CFR Part 54 discussed in Article 14.

A number of other NRC actions on siting and site suitability require environmental reviews, including issuance of limited work authorizations (10 CFR 50.10(e)(1) to (e)(3), 10 CFR 52.25, and 10 CFR 52.91), early partial decisions (10 CFR B , Subpart F), and pre-application early reviews of site suitability issues (10 CFR Part 52, Appendix Q).

Guidance in the environmental standard review plan (1) instructs the NRC staff responsible for environmental reviews; (2) describes how the NRC reaches judgments on environmental impacts caused by constructing and operating nuclear power plants, and by allowing for period of extended operation (license renewal) and refurbishment activities; and (3) specifies how to determine the significance of these impacts. In its environmental review, the NRC staff ensures the following:

- Data essential to a specific environmental review and subsequent decisionmaking are provided and reviewed in the applicant's submittal.

- Appropriate consideration, including coordination and consultation, is given to other Federal, State, regional, local, and affected Native American tribal requirements applicable to a particular environmental review.

- The analysis and evaluation procedures for review of a given technical area are standardized, thus achieving uniformity of approach.

- Each impact assessment concentrates on review of the potential environmental impacts of significance, and analysis of irrelevant data or insignificant impacts is minimized.

- The methods to be used for analysis and staff judgments are objective and founded on sound analytical procedures.

17.3.2 Other Considerations for Siting Reviews

The NRC granted the last site approval in 1978. Since then, and coincident with the publication of the initial environmental standard review plan, many changes to the regulatory environment have affected the NRC and applicants seeking site approvals. These include new environmental laws and regulations, changes in policies and procedures resulting from decisions of courts and administrative hearing boards, and changes in the types of authorizations, permits, and licenses issued by the NRC. Some of these changes and their effects on the environmental standard review plans are highlighted in the paragraphs below.

17.3.2.1 Early Site Permits, Standard Design Certifications, and Combined Licenses

In the late 1980s, the NRC issued regulations that provided an alternative licensing framework to 10 CFR Part 50, "Domestic Licensing of Production and Utilization Facilities," which provided for a construction permit followed by an operating license. The new framework provided in 10 CFR Part 52, "Early Site Permits; Standard Design Certifications; and Combined Licenses for Nuclear Power Plants," introduced the concept of approving designs independent of sites, and approving sites independent of designs, and then efficiently linked the approvals to result in the approval to construct and operate the facility.

17.3.2.2 Environmental Justice

Presidential Executive Order 12898, issued in February 1994, instructed Federal agencies to make "environmental justice" part of each agency's mission by addressing disproportionately high and adverse human health or environmental effects of Federal programs, policies, and activities on minority and low-income populations. Although the order was not binding on independent agencies, the NRC agreed to implement the Executive Order to the extent practicable.

17.3.2.3 Yellow Creek Decision

The authority of the NRC is limited in matters that are expressly assigned to the Environmental Protection Agency (EPA) as shown by the decision on Yellow Creek, a Tennessee Valley facility, in 1978. Specifically, the decision determined that the NRC's authority is limited for those matters that are expressly assigned to the EPA by the Federal Water Pollution Control Act Amendments of 1972.

17.3.2.4 Open Access to Transmission Lines and Economic Deregulation

Recent changes in the economic regulation of utilities have expanded the options to be addressed in considering the need for power in environmental impact statements, as required by 10 CFR Part 51, Appendix A(4). Regulatory agencies in some States have initiated economic deregulation, and the Federal Energy Regulatory Commission has adopted regulations to ensure that all power generators have open access to power transmission facilities. The effects of these changes on environmental review procedures are likely to be significant, especially in defining demand, service areas, and benefits.

17.3.2.5 Severe Accident Mitigation Design Alternatives

When the NRC published the original environmental standard review plans, environmental impact statements did not consider design alternatives to mitigate the consequences of severe accidents. Current NRC policy requires consideration of such alternatives in the environmental impact statements that are prepared for an operating license or for license renewal. Severe accident mitigation design alternatives have been included in final environmental statements for the Limerick 1 and 2 (NUREG-0974) operating license reviews, in the Watts Bar supplemental final environmental impact statement (NUREG-0498), and in license renewal supplements to the generic environmental impact statement for license renewal of nuclear power plants (NUREG-1437).

ARTICLE 18. DESIGN AND CONSTRUCTION

Each Contracting Party shall take the appropriate steps to ensure that:

(i)   the design and construction of a nuclear installation provides for several reliable levels and methods of protection (defense in depth) against the release of radioactive materials, with a view to preventing the occurrence of accidents and to mitigating their radiological consequences should they occur

(ii)   the technologies incorporated in the design and construction of a nuclear installation are proven by experience or qualified by testing or analysis

(iii) the design of a nuclear installation allows for reliable, stable, and easily manageable operation, with specific consideration of human factors and the man-machine interface

This section explains the defense-in-depth philosophy, and how it is embodied in the general design criteria of U.S. regulations. It explains how applicants meet the defense-in-depth philosophy, and how the NRC reviews applications and conducts inspections before issuing licenses to ensure that this philosophy is implemented in practice. Next, this section discusses measures for ensuring that the applications of technologies are proven by experience or qualified by testing or analysis. Article 14, under "Verification by Analysis, Surveillance, Testing and Inspection," also addresses this obligation. Finally, this section discusses requirements regarding reliable, stable, and easily manageable operation, specifically considering human factors and the man-machine interface. This obligation is also addressed in Article 12, "Human Factors."

18.1 Defense-in-Depth Philosophy

This section explains the defense-in-depth philosophy followed in regulatory practice, and the governing documents and regulatory process relevant to designing and constructing a nuclear power plant. It also discusses relevant experience and examples.

18.1.1 Governing Documents and Process

The defense-in-depth philosophy, as applied in regulatory practice, requires that nuclear plants contain a series of independent, redundant, and diverse safety systems. The physical barriers for defense-in-depth are the fuel matrix, the fuel rod cladding, the primary coolant pressure boundary, and the containment. The levels of protection in defense-in-depth are (1) a conservative design, quality assurance, and safety culture; (2) control of abnormal operation and detection of failures; (3) safety and protection systems; (4) accident management, including containment protection; and (5) emergency preparedness.

The defense-in-depth philosophy is embodied in Appendix A, "General Design Criteria for Nuclear Power Plants," to 10 CFR Part 50. General design criteria cover protection by multiple fission product barriers, protection and reactivity control systems, fluid systems, containment design, and fuel and radioactivity control.

The general design criteria establish the minimum requirements for the principal design criteria, which in turn establish the necessary design, fabrication, construction, testing, and performance requirements for structures, systems, and components that are important to safety. "Important to safety" structures, systems, and components are those that provide reasonable assurance that the facility can be operated without undue risk to the health and safety of the public.

To ensure that a plant is properly designed and built as designed, that proper materials are used in construction, that future design modifications are controlled, and that appropriate maintenance and operational practices are followed, a good quality assurance program is needed. To meet this need, General Design Criterion 1, "Quality Standards and Records," of Appendix A to 10 CFR Part 50 and its implementing regulatory requirements specified in Appendix B to Part 50, "Quality Assurance Criteria for Nuclear Power Plants and Fuel Reprocessing Plants" establish quality assurance requirements for all activities affecting the safety-related functions of the structures, systems, and components.

Pursuant to 10 CFR 50.34, "Contents of Applications; Technical Information," an applicant for a construction permit must present the principal design criteria for a proposed facility in its preliminary safety analysis report. As guidance in writing a safety analysis report, the applicant may use Regulatory Guide 1.70, "Standard Format and Content of Safety Analysis Reports for Nuclear Power Plants." The safety analysis report must also contain design information for the proposed reactor, and comprehensive data on the proposed site. The report must also discuss various hypothetical accident situations and the safety features to prevent accidents or, if they should occur, to mitigate their effects on both the public and the facility's employees. After obtaining a construction permit under 10 CFR Part 50, the applicant must submit a final safety analysis report to support an application for an operating license, unless it submitted the report with the original application. This report gives the details of the final design of the facility, plans for operation, and procedures for coping with emergencies. The preliminary and final safety analysis reports are the principal documents that the applicant provides for the staff to determine whether the proposed plant can be built and operated without undue risk to the health and safety of the public.

The NRC staff reviews safety analysis reports according to NUREG-0800, "Standard Review Plan for the Review of Safety Analysis Reports for Nuclear Power Plants," to ensure that the general design criteria are satisfied. The plan assures the quality and uniformity of staff reviews of applications to construct or operate nuclear power plants and presents well-defined bases for evaluating proposed changes to the scope and requirements of reviews.

The NRC staff reviews each application to determine whether the plant design meets the Commission's regulations (10 CFR Parts 20, 50, 73, and 100). These reviews include, in part, the characteristics of the site, including the surrounding population, seismology, meteorology, geology and hydrology; the nuclear plant design; the anticipated response of the plant to postulated accidents; the plant operations, including the applicant's technical qualifications to operate the plant; radiological effluents; and emergency planning. In addition, each application for a nuclear installation must have a comprehensive environmental report that provides a basis for evaluating the environmental impact of the proposed facility. Regulatory Guide 4.2, "Standard Format and Content for Environmental Reports," guides applicants on writing environmental reports. The NRC staff reviews the environmental reports according to NUREG-1555, "Standard Review Plan for Environmental Reviews of Nuclear Power Plants." This plan guides the staff in developing its environmental impact statement.

In reviewing an applicant's submittal, the NRC staff, supported by technical assistance, conducts its independent technical studies to review certain safety and environmental matters. The staff states its conclusions in an environmental impact statement and a safety evaluation report, which it continually updates until the time that it grants the license. Under the traditional licensing framework in 10 CFR Part 50, the NRC does not issue an operating license until construction is complete and the Commission makes specified findings.

The NRC maintains surveillance over the construction to ensure compliance with the agency's regulations to protect public health and safety and the environment. Before construction, the NRC inspection program focuses on the applicant's establishing and implementing a Quality Assurance (QA) Program and siting activities. Inspections cover QA activities in design, procurement, and planning for fabrication and construction of the facility. The inspections of this phase are listed in Inspection Manual Chapter 2511, "Light-Water Reactor Inspection Program — Pre-Construction Permit Phase."

During construction, inspectors sample the spectrum of the applicant's activities to confirm that the applicant is adhering to the requirements of the construction permit, and is building the plant according to the approved design and applicable codes and standards. Inspectors of construction look for qualified staff, high-quality materials, conformance to the approved design, and a well-formulated and well-implemented QA program. The inspections for this phase are listed in Inspection Manual Chapter 2512, "Light-Water Reactor Inspection Program — Construction Phase."

As the applicant completes construction, it conducts pre-operational testing to demonstrate the operational readiness of the plant and its staff. Inspectors during this phase seek to determine whether the applicant has developed adequate test plans — both to verify that tests are consistent with NRC requirements, and to ascertain whether the plant and its staff are thoroughly prepared for safe operation. Inspectors during this phase review overall test procedures, and examine selected procedures for technical adequacy. They witness and assess selected tests to verify that the tests meet their objectives and are consistent. Inspectors also review the qualifications of operating staff, and verify that operating procedures and QA plans are properly developed and executed. The inspections for this phase are listed in Inspection Manual Chapter 2513, "Light-Water Inspection Program — Pre-Operational Testing and Operational Preparedness Phase."

About 6 months before obtaining the operating license, the applicant begins a startup phase to prepare for fuel loading and "power ascension." At this point, the NRC issues a low-power license, granting the applicant the authority to load fuel into the reactor core and conduct power ascension testing up to 5-percent power. After the applicant satisfactorily completes this testing, usually after 3 to 6 months, the NRC grants a full-power license. Next, the reactor is loaded with fuel, and the startup test program begins. Just as in pre-operational testing, NRC inspectors emphasize test procedures and results. Specifically, inspectors appraise the licensee's management system for startup testing, analyze test procedures, witness tests, and review licensee evaluations of test results. The inspections for this phase are listed in Inspection Manual Chapter 2514, "Light-Water Inspection Program — Startup Testing Phase," and Inspection Manual Chapter 2515," Light-Water Inspection Program — Operations Phase." The NRC continues its inspection program throughout the remaining operating life of the plant.

18.1.2 Experience and Examples

For more than 30 years, the AEC and then the NRC have reviewed applications for operating licenses and documented their reviews in safety evaluation reports and their supplements for 110 nuclear installations. These organizations issued an operating license for every facility at the completion of construction.

18.2 Technologies Proven by Experience or Qualified by Testing or Analysis

The NRC ensures that new technologies are proven as required by paragraph (b) of 10 CFR 52.47, "Contents of Application." This rule requires demonstration of new technologies through analysis, appropriate test programs, experience, or a combination thereof.

The earlier discussion under this article (18.1.1) and Article 14, Section 14.2 address the qualification of currently used technologies.

18.3 Design for Reliable, Stable, and Easily Manageable Operation

The NRC specifically considers human factors and the man-machine interface in the design of nuclear installations. For safety analysis reports, the NRC reviews the human factors engineering design of the main control room and the control centers outside of the main control room. As guidance, the staff uses Chapter 18 of Revision 1 NUREG-0800, and Revision 1 to NUREG-0700, "Human-System Interface Design Review Guideline." The NRC also uses NUREG-0711, "Human Factors Engineering Program Review Model," for design certification reviews that include evaluating the design process as part of the final design of next-generation main control rooms. Human factors are also discussed in Article 12.

ARTICLE 19: OPERATION

Each Contracting Party shall take appropriate steps to ensure that:

(i) the initial authorization to operate a nuclear installation is based upon an appropriate safety analysis and a commissioning programed demonstrating that the installation, as constructed, is consistent with design and safety requirements

(ii) operational limits and conditions derived from the safety analysis, test, and operational experience are defined and revised as necessary for identifying safe boundaries for operation

(iii) operation, maintenance, inspection, and testing of a nuclear installation are conducted in accordance with approved procedures

(iv) procedures are established for responding to anticipated operational occurrences and to accidents

(v) necessary engineering and technical support in all safety related fields is available throughout the lifetime of a nuclear installation

(vi) incidents significant to safety are reported in a timely manner by the holder of the relevant license to the regulatory body

(vii) programs to collect and analyze operating experience are established, the results obtained and the conclusions drawn are acted upon and that existing mechanisms are used to share, important experience with international bodies and with other operating organizations and regulatory bodies

(viii) the generation of radioactive waste resulting from the operation of a nuclear installation is kept to the minimum practicable for the process concerned, both in activity and in volume, and any necessary treatment and storage of spent fuel and waste directly related to the operation and on the same site as that of the nuclear installation take into consideration conditioning and disposal

The NRC relies on regulations in Title 10, "Energy," of the *U.S. Code of Federal Regulations* (10 CFR) and internally developed associated programs in granting the initial authorization to operate a nuclear installation and in monitoring its safe operation throughout its life. The material that follows describes the more significant regulations and programs corresponding to each obligation of Article 19.

19.1 Initial Authorization to Operate
19.1.1 Governing Documents and Process

In the past, the NRC licensed nuclear power plants under the traditional (two-step) licensing framework of 10 CFR Part 50. This framework requires both a construction permit and an operating license. The alternative licensing framework 10 CFR Part 52, "Early Site Permits; Standard Design Certifications; and Combined Licenses for Nuclear Power Plants," provides for site approvals and design approvals in advance of construction. In addition, this framework affords a process that combines a construction permit and an operating license (with conditions) into one license. Both frameworks require NRC approval to construct and operate a nuclear power plant.

In addition, each application to construct or operate a nuclear power plant receives an independent review by the Advisory Committee on Reactor Safeguards, an independent statutory committee established to advise the NRC on reactor safety. The Committee begins its review early in the licensing process by selecting proper stages at which to hold a series of meetings with the applicant and the NRC staff. Upon completing its review, the Committee reports to the Commission.

The public also has an opportunity to have any concerns addressed. The Atomic Energy Act requires that a public hearing be held before a construction permit, early site permit, or a combined license may be issued for a nuclear power plant. The public hearing is conducted by a three-member Atomic Safety and Licensing Board, which consists of one lawyer who acts as chairperson, and two technically qualified persons. Members of the public may submit statements to the licensing board, or they may petition for leave to intervene as full parties in the hearing.

To obtain NRC approval to construct or operate a nuclear power plant, an applicant must submit safety analysis reports. Article 18 describes safety reports and reviews that apply to the issuance of an operating license. A public hearing is neither mandatory nor automatic for an application for an operating license under 10 CFR Part 50. However, soon after the NRC accepts the application for review, it publishes a notice that it is considering issuing the license. This notice states that any person whose interest might be affected by the proceeding may petition the NRC for a hearing. If a public hearing is held, the same process described for the hearing for the construction permit applies.

A combined license, issued under Subpart C, "Combined Licenses," of 10 CFR Part 52, authorizes construction of a facility in a manner similar to a construction permit under 10 CFR Part 50. Just as for a construction permit, a hearing must be held before the issuing of a combined operating license. However, the combined license will specify the inspections, tests, and analyses that the licensee must perform and the acceptance criteria that, if met, are necessary and sufficient to provide reasonable assurance that the facility has been constructed and will be operated in conformity with the license and the applicable regulations. After issuing a combined license, the NRC staff will verify that the required inspections, tests, and analyses have been performed and, before operation of the facility, must find that the acceptance criteria have been met. At periodic intervals during construction, the NRC staff will publish notices of the successful completion of inspections, tests, and analyses in the *Federal Register*. Then, not less than 180 days before the date scheduled for initial loading of fuel, the NRC will publish a notice of intended operation of the facility in the *Federal Register*. An opportunity for hearing exists after construction, but petitions for a hearing will only be considered if the petitioner demonstrates that the acceptance criteria have not been met.

An early site permit, issued under Subpart A of 10 CFR Part 52, provides for resolution of site safety, environmental protection, and emergency preparedness issues, independent of a specific nuclear plant review. The application for an early site permit must address the safety and environmental characteristics of the site, and evaluate potential physical impediments to the development of an emergency plan. (Additional detail may be submitted on emergency preparedness issues up to a complete emergency plan.) The staff documents its findings on site safety characteristics and emergency planning in a safety evaluation report, and findings on environmental protection issues in an environmental impact statement. The early site permit also has provisions to perform non-safety site preparation activities, subject to redress, before the issuance of a combined license. After the NRC staff and the Advisory Committee on Reactor Safeguards complete the safety reviews, the NRC will issue a *Federal Register* notice for a mandatory public hearing. The early site permit is valid for no less than 10, nor more than 20, years and can be renewed for 10 to 20 years.

The NRC may certify and approve a standard plant design through a rulemaking, independent of a specific site. The design certification is valid for 15 years, and may be renewed. The issues resolved in a design certification have a more restrictive backfit requirement than issues resolved under other licenses. That is, a certified design cannot be modified by the NRC unless the modification is necessary to meet the applicable regulations in effect during design certification, or to ensure adequate protection of public health and safety. An application for a combined license under 10 CFR Part 52 can incorporate by reference a design certification or an early site permit or both. The advantage of this approach is that the issues resolved by rulemaking for design certification and those resolved during the early site permit hearing process are precluded from reconsideration at the combined license stage.

19.1.2 Experience and Examples

All currently operating reactors were licensed under 10 CFR Part 50. In each case, the NRC scheduled its review of an operating licensing application for 3 years but the actual timing varied, depending on such factors as the completeness of necessary information by the license applicant, delays in construction or resolution of safety issues, and the duration of the public hearing. For the Millstone Unit 3 operating license, no public hearing was requested, and the operating license review was performed in 3 years. For the Comanche Peak operating license, resolution of substantial construction quality issues extended the time required to complete the operating license review.

19.2 Operational Limits and Conditions Are Defined and Revised

The license for each nuclear facility must contain technical specifications that set operational limits and conditions derived from the safety analyses, tests, and operational experience. 10 CFR 50.36, "Technical Specifications," states the requirements that apply to the plant-specific technical specifications. At a minimum, the technical specifications must describe the specific characteristics of the facility, and the conditions for its operation that are required to adequately protect the health and safety of the public. Each applicant is required to identify items that directly apply to maintaining the integrity of the physical barriers that are designed to contain radioactive material. Specifically, 10 CFR 50.36 requires that the technical specifications must be derived from the analyses and evaluation in the safety analysis report. Licensees cannot change the technical specifications without prior NRC approval.

In 1992, the NRC issued improved vendor-specific standard technical specifications and published them in the following reports:

- NUREG-1430, "Standard Technical Specifications, Babcock and Wilcox Plants"
- NUREG-1431, "Standard Technical Specifications, Westinghouse Plants"
- NUREG-1432, "Standard Technical Specifications, Combustion Engineering Plants"
- NUREG-1433, "Standard Technical Specifications, General Electric Plants, BWR/4"
- NUREG-1434, "Standard Technical Specifications, General Electric Plants, BWR/6"

In 1993, the NRC published a "Final Policy Statement on Technical Specifications Improvements for Nuclear Power Reactors." This policy statement identified four criteria to determine what limiting conditions of operation should be defined in technical specifications. A technical specification must be established for each item meeting one or more of the criteria below:

(1) installed instrumentation used to detect, and indicate in the control room, a significant abnormal degradation of the reactor coolant pressure boundary

(2) a process variable, design feature, or operating restriction that is an initial condition of a design-basis accident or transient analysis that either assumes the failure of, or presents a challenge to, the integrity of a fission product barrier

(3) structure, system, or component that is part of the primary success path and which functions or actuates to mitigate a design-basis accident or transient that either assumes the failure of, or presents a challenge to, the integrity of a fission product barrier

(4) a structure, system, or component that operating experience or probabilistic risk assessment has shown to be significant to public health and safety

In 1995, the NRC published a final rule that codified the criteria for determining the content of the technical specifications, and amended 10 CFR 50.36 to include the four criteria.

The NRC periodically revises the standard technical specification NUREGs on the basis of experience. The review and approval of a proposed generic change to the standard technical specifications are multi-stage processes. These processes are designed to ensure that each technical specification remains internally consistent, that consistency is maintained among NUREGs, and that the knowledge and positions of the industry and the NRC are incorporated. In April 1995, the NRC issued Revision 1 to the improved standard technical specifications. The agency issued Revision 2 in June 2001.

The NRC encourages licensees to use the improved standard technical specifications as the basis for plant-specific technical specifications. The NRC will also consider requests to adopt parts of the improved standard technical specifications, even if the licensee does not adopt all of the improvements. These parts will include all related requirements, and will normally be developed as line-item improvements. To date, 62 operating commercial nuclear plants have converted their technical specifications to the improved standard technical specifications.

A licensee may propose re-locating the limiting conditions for operation that do not meet any of the criteria and their associated actions and surveillance requirements from technical specifications to licensee-controlled documents, such as the final safety analysis report. In such cases, the change is processed as a typical license amendment request.

19.3 Approved Procedures

In the United States, operations, maintenance, inspection, and testing of a nuclear installation are conducted in accordance with approved procedures. Each nuclear facility is required to follow the quality assurance requirements in Appendix B to 10 CFR Part 50, "Quality Assurance Criteria for Nuclear Power Plants and Fuel Reprocessing Plants." The Quality Assurance Program is described in Article 13. Criterion V, "Instructions, Procedures, and Drawings," to Appendix B of 10 CFR Part 50, requires that licensees must establish measures to ensure that activities that affect quality will be prescribed by appropriate documented instructions, procedures, or drawings. Activities that affect quality include operation, maintenance, inspection, and testing of the facility.

Revision 3 to NRC Regulatory Guide 1.33, "Quality Assurance Program Requirements (Operations)," gives supplemental guidance. An appendix to this regulatory guide lists specific activities that should be covered by written procedures, including administrative procedures; general plant procedures; operating procedures; procedures for startup, operation, and shutdown of safety-related systems (for both pressurized-water reactors and boiling-water reactors); procedures for dealing with emergencies and other significant events; procedures for controlling radioactivity procedures; for measuring and test equipment; procedures for surveillance tests, inspections, calibrations, and maintenance; and chemical and radiochemical control.

The rule that addresses the need to perform maintenance according to approved procedures is 10 CFR 50.65, "Requirements for Monitoring the Effectiveness of Maintenance at Nuclear Power Plants." The main aim of this rule is to ensure that nuclear power plants will be adequately maintained.

The NRC has changed the rule to add a new section as 10 CFR 50.65(a)(4), which specifically requires licensees to assess and manage the increase in risk that may result from proposed maintenance activities. The need for the rule change arose from licensee efforts to improve overall plant capacity factor by increasing both the amount and frequency of maintenance performed during power operation. Issued on July 19, 1999, the revised rule became effective on November 28, 2000, following the issuance of Regulatory Guide 1.182, "Assessing and Managing Risk Before Maintenance Activities at Nuclear Power Plants."

The benefits of performing maintenance activities during power operations include increased system and plant reliability, reduction of plant equipment and system material condition deficiencies that could adversely impact plant operations, and reduction of work scope during plant refueling outages. However, relevant margins of safety could be inadvertently reduced under certain conditions, for example, if maintenance is performed at power without proper controls and careful consideration of risk. The intent of 10 CFR 50.65(a)(4) is to require that licensees perform assessments before maintenance activities are performed on structures, systems, and components covered by the Maintenance Rule and manage the increase in risk that may result from the proposed activities. The results of these assessments are to be used in conjunction with other regulatory requirements and, therefore, cannot be used as justification to perform activities which may not comply with other regulations.

Licensees may elect any suitable methods or approaches to carry out the rule. The rule does not address the many industry programs that are currently in place to improve maintenance, and may be used in implementing the rule. For example, the rule does not discuss work planning and scheduling, preventive and corrective maintenance procedures, training, post-maintenance testing, work history, methods to determine cause.

The rule is intended to maximize the use of existing industry programs, studies, initiatives, and databases. A licensee can use the results of existing programs to support the demonstration that it is effectively controlling performance of systems, structures, and components by preventive maintenance. If monitoring indicates that performance is unacceptable, a determination of cause should correct any equipment or program deficiency.

Since the industry had limited experience with risk-informed, performance-based regulation, it initially had some widespread problems in implementing the rule. As a result, the NRC decided to inspect programs and performance at each nuclear facility. Once inspectors completed baseline inspections of all plants, they focused routine inspections on equipment performance, and only evaluated programs and processes if causes of problems were rooted in them.

19.4 Procedures for Responding to Anticipated Operational Occurrences and Accidents

This section discusses the documents providing recommendations and guidance on procedures for responding to anticipated operational occurrences and accidents. These are NUREG-0737, "Clarification of TMI Action Plan Requirements," NUREG-0737, Supp. 1, "Requirements for Emergency Response Capability," and NUREG-0899, "Guidelines for the Preparation of Emergency Operating Procedures."

After the accident at TMI-2, in 1979, the NRC issued orders requiring licensees to develop procedures for coping with certain plant transients and postulated accidents. It also issued NUREG-0737 in 1980, and Supplement 1 in 1983. These two documents recommend that licensees should develop procedures to cope with accidents and transients that are caused by initiating events analyzed in the final safety analysis report with multiple failures of equipment. If such failures were unmitigated, conditions of inadequate core cooling would exist. Examples of multiple failure events are (1) multiple tube ruptures in a single steam generator and tube ruptures in more than one steam generator, and (2) failure of main and auxiliary feedwater systems.

NUREG-0899 gives programmatic guidance for developing emergency operating procedures. To aid the NRC's review of emergency operating procedures, NUREG-0899 cites four aspects as providing an adequate basis for a review. These are (1) plant-specific technical guidelines, (2) a plant-specific writer's guide, (3) a description of the program for verifying and validating the procedures, and (4) a description of the program for training operators in the procedures. To ensure that proper procedures had been developed to respond to plant transients and accidents, the NRC reviewed each plant using the guidance in NUREG-0800, Section 13.5.2, "Operating and Maintenance Procedures," which focused on ensuring that the licensee's process to develop the procedures was sound and well documented.

19.5 Availability of Engineering and Technical Support

The NRC's Revised Reactor Oversight Process, discussed in Article 6, includes techniques to ensure that adequate engineering and technical support is available throughout the lifetime of a nuclear installation. Several of the inspection procedures focus on ensuring that adequate support programs are maintained. Licensees also report performance indicators. Depending on findings, the NRC conducts additional inspections to focus upon the causes of the performance problems.

19.6 Incident Reporting

Requirements for incident reporting are specified in 10 CFR 50.72, "Immediate Notification Requirements for Operating Nuclear Power Reactors," and in 10 CFR 50.73, "Licensee Event Report System." The NRC modified these rules in 1992 to delete reporting requirements for some events that were determined to be of little or no safety significance. The agency modified the rules again in October 2000 to reduce or eliminate the unnecessary reporting burden for events that have little or no safety significance. The modified rules continue to provide the Commission with reporting of significant events for which the NRC may need to act to maintain or improve reactor safety or to respond to heightened public concern. The modified rules also better align requirements on event reporting with the type of information that the NRC needs to carry out its safety mission. The NRC issued NUREG-1022, Revision 2, "Event Reporting Guidelines, 10 CFR 50.72 and 50.73," concurrently with the rule changes.

Event reporting under these rules since 1984 has contributed significantly to focusing the attention of the NRC and the nuclear industry on the lessons learned from operating experience to improve reactor safety. Over the years, decreasing trends in the number of reactor transients and significant events and improvements in reactor safety system performance have been evident.

19.7 Programs to Collect and Analyze Operating Experience

Operational safety data are reported to or identified by the NRC in event reports; inspection reports; component failure reports; industry reports; reports on operational, safeguards, and security events; reports submitted under 10 CFR Part 21, "Reporting of Defects and Noncompliances;" and reports of operational experience at foreign facilities. The NRC staff screens operational safety data for safety significance and generic implications. The staff responsible for generic communications systematically assesses and screens all nuclear power reactor-related events, reports, and data to determine their significance and need for further action. The staff also (1) develops, coordinates, and issues generic communications, such as regulatory issue summaries, generic letters, bulletins, and information notices to alert the industry to safety concerns that are identified as a result of power reactor events and conditions; (2) identifies the need for an Augmented Inspection Team or Incident Investigation Team response, and coordinates the NRC's participation in establishing the teams; (3) coordinates briefings of operating events, and serves as a focal point for interfacing with the NRC's regional offices and the industry for incoming reports; (4) maintains and administers an "on-call emergency officer" roster, and staffs the daytime emergency officer functions; and (5) develops and conducts programs for major team inspections at licensee facilities that require a number of engineering and operational specialties.

A group of NRC experts in event evaluation, risk assessment, and human factors reviews issues that have potentially generic implications. On the basis of these reviews, the NRC responds by further analyzing the issue, preparing a generic communication, or simply closing out the issue. Typically, this group follows up on about 140 events per year.

Feedback of operational experience consists of carrying out the actions identified by analysis to maintain or improve licensees' safety and safeguards activities and NRC regulatory programs. Followup measures may include collecting additional relevant information, and recommending immediate or long-term changes. Specific followup action may involve changing facility operations or procedures; modifying facility components, systems, or structures; improving operator or staff training; changing regulations, regulatory guides, licensing review procedures or criteria, the inspection program, and research or risk assessment activities; or issuing a generic communication.

The staff uses the Accident Sequence Precursor Program, described in Article 6, to analyze events using probabilistic risk assessment techniques to determine conditional core damage probabilities. This program quantitatively evaluates operational experience, and serves as one of several tools to ensure that important operating lessons are not overlooked.

The staff uses the Revised Reactor Oversight Process to analyze risk significant events or conditions to ensure that plants are operated within prescribed safety limits.

19.8 Radioactive Waste

The NRC has a formal policy on reducing waste. In 1981, it published a formal "Policy Statement on Low-Level Waste Volume Reduction," which addressed the need for waste generators to minimize the amount of waste that they produce, and stated that the NRC will take expeditious action on requests for licensing systems that reduce the volume of waste. Because of the uncertainty concerning the future availability of disposal facilities, the NRC adopted this policy to facilitate continued availability of disposal and, in case storage became necessary for licensees, to mitigate the effects of having to store waste.

In 1981, the NRC published Generic Letter 81-38, "Storage of Low-Level Radioactive Wastes at Power Reactor Sites," which stated that "Some licensees are considering the installation of major volume reduction processes (e.g., incineration, dehydration, or crystallization) to substantially reduce the volume of waste for disposal. You are encouraged to examine the costs and benefits of such processes for your operations."

In 1989, the NRC published Information Notice 89-13, "Alternative Waste Management Procedures in Case of Denial of Access to Low-Level Waste Disposal Sites," which noted that the following actions help to minimize waste:

- Stop unnecessary work that generates waste

- Change processes, procedures, or radionuclides to reduce the volume of generated waste

- Use volume reduction techniques, such as compaction

Reducing the activity of low-level waste presents additional challenges, because metals become activated as a result of their close proximity to the fuel. For fission products, however, the amount (curies) considered to be "low-level waste" is determined by the quantity that leaks from the fuel rods and is, therefore, a function of the fuel rod failures. The current rate of failures is very low. However, keeping the rate of fuel rod failures low is extremely important, and has led to reducing the generation of low-level waste.

Notwithstanding the preceding guidance, the economics of waste disposal in the U.S. have caused practices to minimize waste to flourish, especially for nuclear power reactor licensees. In the past 10 years, disposal costs have risen by a factor of about 10, and volumes of waste produced have decreased by roughly the same amount. Nuclear power reactors now generate only small amounts (1000–2000 cubic feet) of operational waste each year.

In the U.S., waste is put into a form for storage that is both stable and safe, and minimizes the likelihood that it will migrate (if it were a liquid, for example). Waste put into storage is in a form that is suitable for disposal, or at least a form that can be made suitable for disposal.

APPENDIX A: REFERENCES

American National Standards Institute (ANSI), ANSI/ANS 3.1,"Selection, Qualification and Training of Personnel for Nuclear Power Plants," 1981

————, ANSI 18,1, "Selection and Training of Nuclear Power Plant Personnel," 1971

————, ANSI N18.7, "Administrative Controls and Quality Assurance for the Operational Phase of Nuclear Power Plants," 1976

————, ANSI N45.2, "Quality Assurance Program Requirements for Nuclear Power Plants," 1974

————, ANSI N45.2.11, "Quality Assurance Requirements for the Design of Nuclear Power Plants," 1974

————, ANSI N45.2.13, "Quality Assurance Requirements for Control of Procurement of Items and Services for Nuclear Power Plants," 1976

American Nuclear Society (ANS), N18.7 (See ANSI N18.7 above.)

American Society of Mechanical Engineers (ASME), "Boiler and Pressure Vessel Code," Section XI

————, NQA-1, "Quality Assurance Requirements for Nuclear Facilities," 1979

Electric Power Research Institute (EPRI), "Probabilistic Safety Assessment (PSA) Applications Guide," EPRI-TR-105396, August 1995

Federal Emergency Management Agency (FEMA), FEMA-REP-1 (See NUREG-0654.)

————, "Memorandum of Understanding Between Federal Emergency Management Agency and NRC," 58 FR 47996, September 14, 1993

————, Title 44, "Emergency Management and Assistance," *Code of Federal Regulations,* Part 350, "Review and Approval of State and Local Radiological Emergency Plans and Preparedness"

International Atomic Energy Agency, TECDOC-953, "Method for the Development of Emergency Response Preparedness for Nuclear or Radiological Accidents," Vienna, 1997

————, TECDOC-955, "Generic Assessment Procedures for Determining Protective Actions During a Reactor Accident," Vienna, 1997

International Commission on Radiological Protection, ICRP-26, "Recommendations of the International Commission on Radiological Protection (Adopted January 17, 1977)," Oxford: Pergamon Press, 1991

————, ICRP-30, "Limits of Intakes of Radionuclides by Workers," 8 volumes, Oxford, Pergamon Press, 1978–1982

International Nuclear Safety Advisory Group (INSAG), INSAG-3, "Basic Safety Principles for Nuclear Power Plants," 1988

National Council On Radiation Protection and Measurements, NCRP Report No. 91, "Recommendations On Limits For Exposure to Ionizing Radiation," June 1987

National Security Council, Presidential Directive 39, "U.S. Policy on Counterterrorism," June 21, 1995

Nuclear Energy Institute (NEI),

————, NEI 96-07, "Revision 1 Guidelines for 10 CFR 50.59 Implementation," November 2000

————, NEI 97-04, "Nuclear Energy Inst Design Bases Program Guidelines," September 1997

————, NEI 98-03, "Guidelines for Updating FSARs (Final Safety Analysis Reports)," October 1998

Nuclear Management and Resources Council, NUMARC/NESP-007, "Emergency Planning and Preparedness of Nuclear Power Plants," Rev. 2, January 1992

U.S. Atomic Energy Commission,

————, TID-14844, "Calculation of Distance Factors for Power and Test Reactor Sites, 1962"

————, WASH-1258, "Final Environmental Statement Concerning Proposed Rulemaking Action: Numerical Guides for Design Objectives for Operation to Meet the Criterion 'As Low As Practicable' for Radioactive Material in Light-Water-Cooled Nuclear Power Plant Effluents," 3 Volumes," July 1973

U.S. Congress, Administrative Orders Review Act, 28 U.S.C. 2341 et seq.

———, Administrative Procedure Act, 5 U.S.C. 551 et seq.

———, Atomic Energy Act of 1954, as amended, 42 U.S.C. 2011 et seq.

———, Chief Financial Officers Act of 1990, 5 U.S.C. 5313 to 5315 et seq.

———, Clean Air Act, 42 U.S.C. 7401, et seq.

———, Clean Water Act (see Federal Water Pollution Control Act), 33 U.S.C. 1251, et seq.

———, Coastal Zone Management Act, 16 U.S.C.1451 et seq.

———, Comprehensive Environmental Response, Compensation, and Liability Act (CERCLA) of 1980, 42 U.S.C. 9601 et seq.

———, Debt Collection Improvement Act, 5 U.S.C. 5514 et seq.

———, Endangered Species Act, 7 U.S.C. 136 et seq.

———, Energy Policy Act of 1992, 16 U.S.C. 797 note et seq.

———, Energy Reorganization Act of 1974, as amended, 42 U.S.C. 5801 et seq.

———, Federal Advisory Committee Act, 5 U.S.C. App 2 et seq.

———, Federal Water Pollution Control Act (FWPCA), as amended, 33 U.S.C. 1251 et seq.

———, Freedom of Information Act, 5 U.S.C. 552 et seq.

———, Government in the Sunshine Act, 5 U.S.C. 551 et seq.

———, Government Paperwork Elimination Act, 44 U.S.C. 3504n et seq.

———, Hobbs Act (See Administrative Orders Review Act.)

———, Inspector General Act of 1978, 5 U.S.C. 5315 et seq.

———, Low Level Radioactive Waste Policy Amendments Act of 1955, 42 U.S.C. 2021b et seq.

———, National Environmental Policy Act of 1969, as amended, 42 U.S.C. 4321 et seq.

———, National Historic Preservation Act, 16 U.S.C. 470 et seq.

———, Nuclear Non-Proliferation Act of 1978, 22 U.S.C. 3201 et seq.

———, Nuclear Waste Policy Act of 1982, 42 U.S.C. 10101 et seq.

————, Omnibus Budget Reconciliation Act of 1990 (OBRA-90), 20 U.S.C. 1088 et seq.

————, Price-Anderson Act of 1957, 42 U.S.C. 2012 et seq.

————, Privacy Act, 5 U.S.C. 552a et seq.

————, Resource Conservation and Recovery Act (RCRA) of 1976, 42 U.S.C. 6901 et seq.

————, Toxic Substances Control Act of 1976, 15 U.S.C. 2601 et seq.

————, Uranium Mill Tailings Radiation Control Act of 1978, 42 U.S.C. 6907 et seq.

————, West Valley Demonstration Act of 1980, 42 U.S.C 2021a note

————, Wild and Scenic Rivers Act, 16 U.S.C. 1271 et seq.

U.S. Environmental Protection Agency, EPA-400-R-92-001, "Manual of Protective Action Guides and Protective Actions for Nuclear Incidents," May 1992

————, EPA 520/1-78-016 (See NUREG-0396.)

————, EPA-520/1-88-020, "Limiting Values of Radio nuclide Intake and Air Concentration and Dose Conversion Factors for Inhalation, Submersion, and Ingestion, 1988

U.S. Federal Energy Regulatory Commission, "Promoting Wholesale Competition Through Open Access Nondiscriminatory Transmission Services by Public Utilities; Recovery of Stranded Costs by Public Utilities and Transmitting Utilities," 61 FR 21540, May 10, 1996

U.S. Nuclear Regulatory Commission

————, "Agreement Between the Government of the United States of America and the Government of Canada on Cooperation in Comprehensive Civil Emergency Planning and Management" and "Administrative Arrangement Between the United States Nuclear Regulatory Commission and the Atomic Energy Control Board of Canada for Cooperation and the Exchange of Information in Nuclear Regulatory Matters," June 21, 1989

————, "Agreement for the Exchange of Information and Cooperation in Nuclear Safety Matters" and "Implementing Procedure for the Exchange of Technical Information and Cooperation in Nuclear Safety Matters Between The Nuclear Regulatory Commission of the United States of America and the Comision Nacional de Seguridad Nuclear y Salvaguardias of Mexico," October 6, 1989

————, Branch Technical Positions (See NUREG-0800.)

————, "Consideration of Potassium Iodide in Emergency Plans," 66 FR 5247, January 19, 2001

————, DG-1064, An Approach for Plant-Specific, Risk-Informed Decisionmaking: Graded Quality Assurance, June 1997

———, Executive Order 12866, "Regulatory Planning and Review," 58 FR 51735, September 30, 1993

———, Executive Order 12898, "Federal Actions To Address Environmental Justice in Minority and Low-Income Populations," 59 FR 7629, February 1994

———, "Final Policy Statement on the Restructuring and Economic Deregulation of the Electric Utility Industry," 62 FR 44071, August 19, 1997

———, "Final Policy Statement on Technical Specification Improvement for Nuclear Power Reactors," 58 FR 39132, July 22, 1993

———, "Financial Assurance Requirements for Decommissioning Nuclear Power Plants, 63 FR 50465, September 22, 1998

———, "General Statement of Policy and Procedure for NRC Enforcement Actions," 65 FR 250368, May 1, 2000, 65 FR 59274, 65 FR 79139, December 18, 2000

———, Generic Letter 81-38, "Storage of Low-Level Waste at Power Reactor Sites," November 10, 1981

———, Generic Letter 82-02, "Nuclear Plant Staff Working Hours," February 8, 1982

———, Generic Letter 82-12, "Nuclear Power Plant Staff Working Hours," June 15, 1982

———, Generic Letter 82-16, "NUREG-0737 Technical Specifications," September 20, 1982

———, Generic Letter 82-33, "Requirements for Emergency Response Capability," December 17, 1982

———, Generic Letter 83-02, "NUREG-0737 Technical Specifications," January 10, 1983

———, Generic Letter 83-14, "Definition of Key Maintenance Personnel (Clarification of Generic Letter 82-12)," March 7, 1983

———, Generic Letter 85-06, "Quality Assurance Guidance for ATWS Equipment That Is Not Safety Related," April 16, 1985

———, Generic Letter 88-20, "Individual Plant Examination for Severe Accident Vulnerabilities" - 10 CFR 50.54(f)," November 23, 1988

———, Generic Letter 91-18, Rev. 1, "Information to Licensees Regarding Two NRC Inspection Manual Sections on Resolution of Degraded and Nonconforming Conditions and on Operability," November 7, 1991

———, Information Notice 89-13, "Alternative Waste Management Procedures in Case of Denial of Access to Low-Level Waste Disposal Sites," February 8, 1989

———, Information Notice 91-36, "Nuclear Power Plant Staff Working Hours," June 10, 1991

———, Information Notice 97-78, "Crediting of Operator Actions in Place of Automatic Actions and Modifications of Operator Actions, Including Response Times," October 23, 1997

———, Inspection Manual Chapter 2511, "Light-Water Reactor Inspection Program— Pre-Construction Permit Phase"

———, Inspection Manual Chapter 2512, "Light-Water Reactor Inspection Program— Construction Phase"

———, Inspection Manual Chapter 2513, "Light-Water Reactor Inspection Program— Pre-Operational Testing and Operational Preparedness Phase"

———, Inspection Manual Chapter 2514, "Light-Water Reactor Inspection Program— Startup Testing Phase"

———, Inspection Manual Chapter 2515, "Light-Water Reactor Inspection Program— Operations Phase"

———, Inspection Procedure Part 9900 Technical Guidance

———, Inspection Procedure 41500, "Training and Qualification Effectiveness,"

———, Inspection Procedure 42001, "Emergency Operating Procedures"

———, Inspection Procedure 42700, "Plant Procedures"

———, Inspection Procedure 71001, "Licensed Operator Requalification Program Evaluation"

———, Inspection Procedure 71111.11, "Licensed Operator Requalification Program"

———, Inspection Procedure 71152, "Identification and Resolution of Problems"

———, Inspection Procedure 71841, "Supplemental Inspection for Human Performance"

———, Inspection Procedure 95003, " Supplemental Inspection of Repetitive Degraded Cornerstones, Multiple Degraded Cornerstones, Multiple Yellow Inputs, or One Red Input"

———, Management Directive and Handbook 4.2, "Administrative Control of Funds"

———, "The Mission of the Office of Enforcement,"< http://www.nrc.gov

———, NUREG-75/087, "Standard Review Plan for the Review of Safety Analysis Reports for Nuclear Power Plants—LWR Edition," September 1975 (now NUREG-0800)

———, NUREG-0325, Rev. 22, "NRC Organization Charts and Functional Statements," November 1997

———, NUREG-0371, "Approved Category A Task Action Plans," November 1977

———— NUREG- 0396, "Planning Basis for the Development of State and Local Government Radiological Emergency Response Plans in Support of Light Water Nuclear Power Plants, EPA-520/1-78/016," December 1978

————, NUREG-0471, "Generic Task Problem Description, Category B, C, and D Tasks," June 1978

————, NUREG-0654, Rev. 1, "Criteria for Preparation and Evaluation of Radiological Emergency Response Plans and Preparedness in Support of Nuclear Power Plants, " FEMA-REP-1, November 1980

————, NUREG-0654, Rev. 1, Supp. 3, "Criteria for Protective Action Recommendations for Severe Accidents (Draft Report for Interim Use and Comment)," July 1996

————, NUREG-0660, "NRC Action Plan Developed as a Result of the TMI-2 Accident," May 1980

————, NUREG-0700, "Guidelines for Control Room Design Reviews," August 1981

————, NUREG-0700, Rev. 1, "Human-System Interface Design Review Guideline," June 1996

————, NUREG-0711, "Human Factors Engineering Program Review Model," July 1994

————, NUREG- 0713, Vol. 21," Occupational Radiation Exposure at Commercial Nuclear Power Reactors and Other Facilities," October 2000

————, NUREG-0728, Rev. 2, "NRC Incident Response Plan," June 1987

————, NUREG-0737, "Clarification of TMI Action Plan Requirements," November 1980

————, NUREG-0737, Supp. 1, "Requirements for Emergency Response Capability," January 1983

————, NUREG-0761, "Contents of Radiation Protection Plans for Nuclear Power Reactor Licensees," March 1981

————, NUREG-0800, "Standard Review Plan for the Review of Safety Analysis Reports for Nuclear Power Plants," 1981, 1984, and 1987 (formerly NUREG-75/087)

————, NUREG-0899, "Guidelines for the Preparation of Emergency Operating Procedures," August 1982

————, NUREG-0933, Rev. 24, "Program for the Resolution of Generic Issues Related to Nuclear Power Plants," June 2000 <http://www.nrc.gov>

————, NUREG-0974, "Final Environmental Statement: Limerick 1 and 2 Operating License," 1989

————, NUREG-0985, Rev. 2 "Nuclear Regulatory Commission Human Factors Program Plan," April 1986

————, NUREG-1021, Rev. 8, "Operator Licensing Examiner Standards," April 1999

————, NUREG-1021, "Operator Licensing Examiner Standards," Revision 8, Supplement 1, April 2001.

————, NUREG-1022, Rev.1, "Event Reporting Guidelines, 10 CFR 50.72 and 50.73," January 1998

————, NUREG-1055, "Improving Quality and the Assurance of Quality in the Design and Construction of Nuclear Power Plants," May 1984

————, NUREG-1100, Vol. 17, "Budget Estimates, Fiscal Year 2002"

————, NUREG-1150, "Severe Accident Risks: An Assessment for Five U.S. Nuclear Power Plants," December 1990

————, NUREG-1210, "Pilot Program: NRC Severe Reactor Accident Incident Response Training Manual," February 1987

————, NUREG-1220, Rev. 1, "Training Review Criteria and Procedures," January 1993

————, NUREG-1251, "NRC Action Plan Developed as a Result of the TMI-2 Accident, Volume 2," April 1989

————, NUREG-1358, "Lessons Learned From the Special Inspection Program for Emergency Operating Procedures Conducted March–October 1988," April 1989

————, NUREG-1430, "Standard Technical Specifications, Babcock and Wilcox Plants, Vols. 1 and 2," April 1995

————, NUREG-1431, "Standard Technical Specifications, Westinghouse Plants, Vols. 1-3", April 1995

————, NUREG-1432, "Standard Technical Specifications, Combustion Engineering Plants, Vols. 1-3," April 1995

————, NUREG-1433, "Standard Technical Specifications, General Electric Plants, BWR/4, Vols. 1-3," April 1995

————, NUREG-1434, "Standard Technical Specifications, General Electric Plants, BWR/6, Vols. 1 and 2," April 1995

————, NUREG-1437, "Generic Environmental Impact Statement for License Renewal of Nuclear Power Plants," May 1996 <http://www.nrc.gov>

————, NUREG-1462, "Final Safety Evaluation Report Related to the Certification of the System 80+ Design," August 1994

————, NUREG-1465, "Accident Source Terms for Light-Water Nuclear Power Plants," February 1995

————, NUREG-1503, "Final Safety Evaluation Report Related to the Certification of the Advanced Boiling Water Reactor Design," July 1994

————, NUREG-1512, "Final Safety Evaluation Report Related to the Certification of the AP 600 Design," November 1994

————, NUREG-1555 and 1555 Supplement 1, "Standard Review Plan for Environmental Reviews of Nuclear Power Plants, Operating License Renewal," March 2000

————, NUREG-1560, "Individual Plant Examination: Perspectives on Reactor Safety and Plant Performance, December" 1997

————, NUREG-1577, "Standard Review Plan on Power Reactor Licensee Financial Qualifications and Decommissioning Funding Assurance," January 1997 <http://www.nrc.gov>

————, NUREG-1600, "General Statement of Policy and Procedures for NRC Enforcement Actions (Enforcement Policy)," July 2000

————, NUREG-1649, Rev. 1 "New NRC Reactor Inspection and Oversight Program," April 2000

————, NUREG-1649, Rev. 3 " Reactor Oversight Process," July 2000 <http://www.nrc.gov/NRC>

———— , NUREG-1742, "Perspectives Gained from the Individual Plant Examination of External Events (IPEEE) Program," October 2001

————, NUREG-1800, "Standard Review Plan for the Review of License Renewal Applications for Nuclear Power Plants," July 2001

————, NUREG-1801, "Generic Aging Lessons Learned (GALL) Report," July 2001

————, NUREG/BR-0058, Revision 3, "Regulatory Analysis Guidelines of the U.S. Nuclear Regulatory Commission," July 2000

————, NUREG/BR-0150, Vol. 1, Rev. 4, "RTM 96—Response Technical Manual," March 1996

————, NUREG/BR-0195, "NRC Enforcement Manual," Rev. 3, 65 FR 25368, May 1, 2000, 65 FR 59274, 65 FR 79139 <http:/www.nrc.gov.>

————, NUREG/BR-0230, "RCM-96, Response Coordination Manual," September 1996

———, NUREG/CR-2601, "Technology, Safety and Costs of Decommissioning Reference Light Water Reactors Following Postulated Accidents," November 1982

———, NUREG/CR-2850, Vol. 14, "Dose Commitments Due to Radioactive Releases From Nuclear Power Plant Sites in 1992," March 1996

———, NUREG/CR-4674, Vol. 27, "Precursors to Potential Severe Core Damage Accidents: 1999—A Status Report," July 2000

———, NUREG/CR-6093, "An Analysis of Operational Experience During Low Power and Shutdown and a Plan for Addressing Human Reliability Assessment Issues," June 1994

———, NUREG/CR-6116, "Systems Analysis Programs for Hands-On Integrated Reliability Evaluations (SAPHIRE) Version 5.0," Vols. 1-10, December 1993 through April 1995

———, NUREG/CR-6617, "Price Anderson Act—Crossing the Bridge to the Next Century: Report to Congress," August 1998 <http://www.nrc.gov>

———, NUREG/CR-6633, "Advanced Information System Design: Technical Basis and Human Factors Review Guidance," March 2000

———, NUREG/CR-6634, "Computer-Based Procedure Systems: Technical Basis and Human Factors Guidance," March 2000

———, NUREG/CR-6635, "Soft Controls: Technical Basis and Human Factors Review Guidance," March 2000

———, NUREG/CR-6636, "Maintainability of Digital Systems: Technical Basis and Human Factors Guidance," March 2000

———, NUREG/CR-6637, "Human Systems Interface and Plant Modernization Process: Technical Basis and Human Factors Review Guidance," March 2000

———, NUREG/CR-6684, "Advanced Alarm Systems: Revision of Guidance and Its Technical Basis," November 2000

———, NUREG/CR-6689, "A Proposed Approach for reviewing Changes to Risk Important Human Actions," October 2000

———, NUREG/CR-6691, "The Effects of Alarm Display, Processing, and Availability on Crew Performance," November 2000

———, NUREG/IA-0137, "A Study of Control Room Staffing Levels for Advanced Reactors," November 2000

———, Office Instruction LIC-100, "Control of Licensing Bases for Operating Reactors," March 2001

————, Office Letter No. 906, Revision 1, "Procedural Guidance for Preparing Environmental Assessments and Considering Environmental Issues," Office of Nuclear Reactor Regulation," 1996

————, "Policy Statement on Engineering Expertise on Shift," 50 FR 43621, October 28, 1985

————, "Policy Statement on Factors Causing Fatigue of Operating Personnel at Nuclear Reactors," 47 FR 23836, June 1, 1982

————, "Policy Statement on Low-Level Waste Reduction," 46 FR 51100, October 16, 1981

————, "Policy Statement on Severe Reactor Accidents Regarding Future Designs and Existing Plants," 50 FR 32138, August 8, 1985

————, "Policy Statement on Nuclear Power Plant Staff Working Hours," 47 FR 23836, June 11, 1982

————, "Policy Statement on Use of PRA Methods in Nuclear Activities," 60 FR 42623, August 16, 1995.

————, "Part 20—Standards for Protection Against Radiation," 56 FR 23391, May 21, 1991

————, RG 1.3, Rev. 2, "Assumptions Used for Evaluating the Potential Radiological Consequences of a Loss of Coolant Accident for Boiling Water Reactors," June 1974

————, RG 1.4, Rev 2, "Assumptions Used for Evaluating the Potential Radiological Consequences of a Loss of Coolant Accident for Pressurized Water Reactors," June 1974

————, RG 1.8, "Personnel Selection and Training," March 1971

————, RG 1.8, Rev. 1, "Personnel Selection and Training," September 1975 (for comment) and May 1977

————, RG 1.8, Rev.2, "Qualification and Training of Personnel for Nuclear Power Plants," April 1987

————, RG 1.28, Rev. 3, "Quality Assurance Program Requirements (Design and Construction)" August 1985

————, RG 1.33, Rev. 2, "Quality Assurance Program Requirements (Operations)," February 1978

————, RG 1.70, Rev. 3, "Standard Format and Content of Safety Analysis Reports for Nuclear Power Plants," November 1978

————, RG 1.101 "Emergency Planning and Preparedness for Nuclear Power Plants," Rev. 3, August 1992

————, RG 1.145, Rev. 1, "Atmospheric Dispersion Models for Potential Accident Consequence Assessments at Nuclear Power Plants," November 1982 (reissued with corrected page 1.145-7, February 1983)

————, RG 1.155, "Station Blackout, June 1988 (reissued with corrected tables," August 1988)

————, RG 1.159, "Assuming the Availability of Funds for Decommissioning Nuclear Reactors," August 1990

————, RG 1.174, "An Approach for Using Probabilistic Risk Assessment in Risk-Informed Decisions on Plant-Specific Changes to the Current Licensing Basis" July 1998

————, RG 1.175, "An Approach for Plant-Specific, Risk-Informed Decisionmaking: Inservice Testing (Draft DG-1062 issued 6/97) (Issued with SRP Chapter 3.9.7)," August 1998 <http://www.nrc.gov>

————, RG 1.177, "An Approach for Plant-Specific, Risk-Informed Decisionmaking: Technical Specifications," August 1998

————, RG 1.178, "An Approach for Plant-Specific Risk-Informed Decisionmaking: Inservice Inspection of Piping," September 1998

————, RG 1.181, "Content of the Updated Final Safety Analysis Report in Accordance with 10 CFR 50.71(e) (Draft DG-1083 issued 3/99)," September 1999 <http://www.nrc.gov>

————, RG 1.182, "Assessing and Managing Risk before Maintenance Activities at Nuclear Power Plants," May 2000

————, RG 1.183, "Alternative Radiological Source Terms for Evaluating Design Basis Accidents at Nuclear Power Plants," July 2000

————, RG 1.186, "Guidance and Examples of Identifying 10 CFR 50.2 Design Bases (Draft DG-1093 published 4/00)," December 2000

————, RG 1.187, "Guidance for Implementation of 10 CFR 50.59, Changes, Tests, and Experiments (Draft DG-1095 published 4/00)," November 2000

————, RG 1.188, "Standard Format and Content of Applications to Renew Nuclear Power Plants Operating Licenses," July 2001

————, RG 4.2, Rev. 2, "Preparation of Environmental Reports for Nuclear Power Stations," July 1976 and "Supplement 1, Preparation of Supplemental Environmental Reports for Applications to Renew Nuclear Power Plants Operating Licenses," August 1991

————, RG 8.8, "Information Relevant to Maintaining Occupational Radiation Exposure As Low As Practicable (Nuclear Reactors), U.S. Atomic Energy Commission," July 1973

————, RG 8.8, Rev.1, "Information Relevant to Maintaining Occupational Radiation Exposure As Low As Is Reasonably Achievable," September 1975

———, RG 8.8, Rev. 2, "Information Relative to Ensuring That Occupational Radiation Exposures at Nuclear Power Stations Will Be As Low As Is Reasonably Achievable," March 1977

———, RG 8.8, Rev. 3, "Information Relative to Ensuring That Occupational Radiation Exposures at Nuclear Power Stations Will Be As Low As Is Reasonably Achievable," June 1978

———, "Safety Goals for the Operation of Nuclear Power Plants: Policy Statement; Republication 51 FR 30028," August 21, 1986

———, SECY-88-147, "Integration Plan for Closure of Severe Accident Issues," May 25, 1988

———, SECY-93-184, "Shift Staffing at Nuclear Plants," June 29, 1993

———, SECY-93-193, "Policy on Shift Technical Advisor Position at Nuclear Power Plants," July 13, 1993

———, SECY-97-229, "Graded Quality Assurance/Probabilistic Risk Assessment Implementation Plan for the South Texas Project Electric Generating Station," October 6, 1997

———, SECY-98-012, "Quarterly Status Report on the Probabilistic Implementation Plan," January 23, 1998

———, SECY-98-015, "Final General Regulatory Guide and Standard Review Plan for Risk-Informed Regulation of Power Reactors," January 30, 1998

———, SECY-98-154, "Results of Revised NUREG-1465 Source Term Rebaselining for Operating Reactors," June 30, 1998

———, SECY-98-279, "Partial Granting of Petition for Rulemaking Submitted by the Nuclear Energy Institute (PRM-50-62)", November 30, 1998

———, SECY-00-053, "NRC Program on Human Performance in Nuclear Power Plant Safety," February 29, 2000

———, SECY-00-213, "Risk-Informed Implementation Plan," October 26, 2000

———, SECY-00-159, "Final Rule Amending the Fitness-for-Duty Rule," December 4, 2000

———, "Severe Accident Policy Statement, 51 FR 32138," August 8, 1985

———, Title 10, "Energy," Code of Federal Regulations <http://www.nrc.gov>

    Part 2, "Rules of Practice for Domestic Licensing Proceedings and Issuance of Orders"

    Part 2, Subpart A, "Procedure for Issuance, Amendment, Transfer or Renewal of a License"

Part 2, Subpart B," Procedure for Imposing Requirements by Order, or for Modification, Suspension, or Revocation of a License, or for Imposing Civil Penalties"

Part 2, Subpart F, "Additional Procedures Applicable to Early Partial Decisions on Site Suitability Issues in Connection With an Application for a Permit to Construct Certain Utilization Facilities"

Part 2, Subpart G, " Rules of General Applicability"

§2.101, "Filing of application"

§2.201, "Notice of Violation"

§2.202, "Orders"

§2.204, "Demand for Information"

§2.205, "Civil Penalties"

Part 19, "Notices, Instructions and Reports to Workers: Inspection and Investigations"

Part 20, "Standards for Protection Against Radiation"

Part 21, "Reporting of Defects and Noncompliances"

Part 26, "Fitness for Duty Programs"

Part 30, "Rules of General Applicability to Domestic Licensing By Product Material"

Part 30, Appendix A, "Criteria Relating to the Use of Financial Tests and Parent Company Guarantees for Providing Reasonable Assurance for Funds for Decommissioning"

Part 30, Appendix C, "Criteria Relating to the Use of Financial Tests and Self Guarantees of Providing Reasonable Assurance for Funds for Decommissioning"

Part 30, Appendix D, "Criteria Relating to the Use of Financial Tests and Self Guarantees of Providing Reasonable Assurance for Funds for Decommissioning of Commercial Companies that have no Outstanding Rated Bonds"

Part 34, "Licenses for Industrial Radiography ;and Radiation Safety Requirements for Industrial Radiographic Operations"

Part 35, "Medical Use of Byproduct Material"

Part 39, "Licenses and Radiation Safety Requirements for Well Logging"

Part 40, "Domestic Licensing of Source Material"

Part 50, "Domestic Licensing of Production and Utilization Facilities"

Part 50, Appendix A, "General Design Criteria for Nuclear Power Plants,"

§50.2, "Definitions"

§50.10, "License required"

§50.33, "Contents of applications; general information"

§50.33, "Contents of applications; general information"

§50.34, "Contents of applications; technical information"

§50.36, "Technical Specifications"

§50.40, "Common Standards"

§50.47, "Emergency Plans"

§50.48, "Fire Protection"

§50.49, "Environmental Qualification of Electrical Equipment Important to Safety of Nuclear Power Plants"

§50.54, "Conditions of Licenses"

§50.59, "Changes, Tests and Experiments"

§50.61, "Fracture Toughness Requirements for Protection Against Pressurized Thermal Shock Events"

§50.62, "Requirements for Reduction of Risk from Anticipated Transient Without Scram Events for Light-Water Cooled Nuclear Power Plants"

§50.63, "Loss of Alternating Current Power"

§50.65, "Requirements for Monitoring the Effectiveness of Maintenance at Nuclear Power Plants" (the Maintenance Rule)

§50.67, "Accident Source Term"

§50.71, "Requirements for Updating Final Safety Analysis Reports"

§50.72, "Immediate Notification Requirements for Operating Nuclear Power Reactors"

§50.73, "Licensee Event Report System"

§50.75, "Reporting and Record Keeping for Decommissioning Planning"

§50.80, "Transfer of Licenses"

§50.82, "Termination of License"

§50.90, "Application for Amendment of License or Construction Permit"

§50.109, "Backfitting"

§50.120, "Training and Qualification of Nuclear Power Plant Personnel,"

Appendix A to Part 50, "General Design Criteria for Nuclear Power Plants"

Appendix B to Part 50, "Quality Assurance Criteria for Nuclear Power Plants and Fuel Reprocessing Plants," Criterion V, "Instructions, Procedures, and Drawings"

Appendix C to Part 50, "A Guide for the Financial Data and Related Information Required to Establish Financial Qualifications for Facility Construction Permits"

Appendix E to Part 50, "Emergency Planning and Preparedness for Protection and Utilization Facilities"

Appendix I to Part 50, "Numerical Guides for Design Objectives and Limiting Conditions for Operations to Meet the Criterion 'As Low As Is Reasonably Achievable' for Radioactive Material in Light-Water-Cooled Nuclear Power Reactor Effluents"

Appendix J to Part 50, "Containment Leakage"

Appendix Q to Part 50, "Pre-Application Early Review of Site Suitability Issues"

Appendix R to Part 50, "Fire Protection Programs for Nuclear Power Facilities Operating Prior to January 1, 1979

Part 51 "Environmental Protection Regulations for Domestic Licensing and Related Regulatory Functions"

Appendix A to Subpart A of Part 51, "Format for Presentation of Material in Environmental Impact Statements"

Appendix B to Subpart A of Part 51, "Environmental Effect of Renewing the Operating License of a Nuclear Power Plant"

Part 52, "Early Site Permits; Standard Design Certifications; and Combined Licenses for Nuclear Power Plants"

Subpart A of Part 52, "Early Site Permits"

Subpart B of Part 52, "Standard Design Certifications"

Subpart C of Part 52, "Combined Licenses "

§52.25, "Extent of Activities Permitted"

§52.77, "Contents of applications; technical information"

§52.91, "Authorization of Conduct Site Activities"

Appendix O to Part 52, "Final Design Approvals"

Appendix Q to Part 52, "Pre-Application Early Review of Site Suitability Issues"

Part 54, "Requirements for Renewal of Operating Licenses for Nuclear Power Plants"

Part 55, "Operator Licensing"

Part 70, "Domestic Licensing of Special Nuclear Material"

Part 71, "Packaging and Transportation of Radioactive Material"

Part 72, "Licensing Requirements for the Independent Storage of Spent Nuclear Fuel and High-Level Radioactive Waste"

Part 73, "Physical Protection of Plants and Materials"

Part 76, "Certification of Gaseous Diffusion Plants"

Part 100, "Reactor Site Criteria"

§100.10, "Factors to be Considered in Evaluating Sites Before January 1, 1997"

§100.20, "Factors to be Considered in Evaluating Sites on or After January 1, 1997"

Appendix A to Part 100, "Seismic and Geologic Siting Criteria for Nuclear Power Plants"

Part 140, "Financial Protection Requirements and Indemnity Agreements"

Part 170, "Fees for Facilities, Materials, Import and Export Licenses and Other Regulatory Services under the Atomic Energy Act of 1954, as Amended"

Part 171, "Annual Fees for Reactor Licenses and Fuel Cycle Licenses and Materials Licenses, Including Holders of Certificates for Compliance, Registrations, and Quality Assurance Program Approvals and Government Agencies Licensed by the NRC"

APPENDIX B: ABBREVIATIONS

AEA	Atomic Energy Act of 1954, as amended
ALARA	as low as is reasonably achievable
ANSI	American National Standards Institute
ASME	American Society of Mechanical Engineers
BEIR	U.S. National Academy of Sciences Committee on the Biological Effects of Ionizing Radiation,
CFR	*U.S. Code of Federal Regulations*
DOE	U.S. Department of Energy
EPA	U.S. Environmental Protection Agency
ERA	Energy Reorganization Act of 1974
ERDA	U.S. Energy Research and Development Administration
FEMA	Federal Emergency Management Agency
FBI	Federal Bureau of Investigation
FDA	Food and Drug Administration
IAEA	International Atomic Energy Agency
ICRP	International Commission on Radiological Protection
NCRP	National Council on Radiation Protection and Measurements
NEPA	National Environmental Policy Act
PRA	probabilistic risk assessment
QA	quality assurance
SAPHIRE	Systems Analysis Programs for Hands-On Integrated Reliability Evaluation
TEDE	total effective dose equivalent
TMI-1	Three Mile Island, Unit 1
TMI-2	Three Mile Island, Unit 2
USC	U.S. Code
UNSCEAR	United Nations Scientific Committee on the Effects of Atomic Radiation

APPENDIX C: ACKNOWLEDGMENTS

Contributors to this report included the technical and regulatory experts at the NRC listed below. The project manager and principal author was Merrilee Banic, of the Office of Nuclear Reactor Regulation.

Dennis Allison
Merrilee Banic
Douglas Coe
Daniel Dorman
Richard Eckenrode
Jack Foster
Charles Hinson
Stephen Hoffman
Falk Kantor
Steve LaVie
Stuart Magruder
Eileen M. McKenna
Renée Pedersen
Roger Pedersen
Trip Rothschild
Juan Peralta
David Skeen
Robert Stransky
Garmon West
Jerry Wilson
Robert Wood
Barry Zalcman

Other contributors included Dr. Frank Congel (Ombudsman) and Nina M. Barnett (Typist).

ANNEX 1: U.S. Commercial Nuclear Power Reactors

SOURCE: NRC and licensee data as compiled by the NRC

RELEVANT ARTICLE: Introduction and Article 6

U.S. Commercial Nuclear Power Reactors

Unit Operating Utility	Type NSSS AE Constructor	Licensed MWt	Net MDC	CP Issued OL Issued Comm. Op Exp. Date	1995-2000* Average Capacity Factors (Percent)
Arkansas Nuclear 1 Entergy Operations, Inc.	PWR-DRYAMB B&W LLP BECH BECH	2568	0836	12/06/1968 05/21/1974 12/19/1974 05/20/2034	81.6 85.6 99.0 82.6 91.7 87.3
Arkansas Nuclear 2 Entergy Operations, Inc.	PWR-DRYAMB COMB CE BECH BECH	2815	0858	12/06/1972 09/01/1978 03/26/1980 07/17/2018	75.6 93.7 92.6 86.9 82.8 69.9
Beaver Valley 1 First Energy	PWR-DRYSUB WEST 3LP S&W S&W	2652	0810	06/26/1970 07/02/1976 10/01/1976 01/29/2016	76.7 80.0 56.3 33.2 86.1 82.7
Beaver Valley 2 First Energy	PWR-DRYSUB WEST 3LP S&W S&W	2652	0820	05/03/1974 08/14/1987 11/17/1987 05/27/2027	84.1 66.2 85.7 16.9 80.1 86.5
Braidwood 1 Exelon	PWR-DRYAMB WEST 4LP S&L CWE	3411	1100	12/31/1975 07/02/1987 07/29/1988 10/17/2026	67.2 70.5 83.9 78.6 101.0 96.4

U.S. Commercial Nuclear Power Reactors (Cont.)

Unit Operating Utility	Type NSSS AE Constructor	Licensed MWt	Net MDC	CP Issued OL Issued Comm. Op Exp. Date	1995-2000* Average Capacity Factors (Percent)
Braidwood 2 Exelon	PWR-DRYAMB WEST 4LP S&L CWE	3411	1100	12/31/1975 05/20/1988 10/17/1988 12/18/2027	97.2 81.3 85.5 97.4 92.0 98.4
Browns Ferry 1 Tennessee Valley Authority	BWR-MARK1 GE 4 TVA TVA	3293	0	05/10/1967 12/20/1973 08/01/1974 12/20/2013	0.0 0.0 0.0 0.0 0.0 0.0
Browns Ferry 2 Tennessee Valley Authority	BWR-MARK1 GE 4 TVA TVA	3293	1065	05/10/1967 08/02/1974 03/01/1975 06/28/2014	98.6 86.0 89.7 98.9 89.1 99.1
Browns Ferry 3 Tennessee Valley Authority	BWR-MARK 1 GE TVA TVA	3293	1065	07/31/1968 08/18/1976 03/01/1977 07/02/2016	70.4 94.1 91.4 80.8 99.4 92.6
Brunswick 1 Carolina Power & Light Co.	BWR-MARK 1 GE 4 UE&C BRRT	2558	0767	02/07/1970 11/12/1976 03/18/1977 09/08/2016	85.9 84.7 102.1 83.6 97.4 93.7
Brunswick 2 Carolina Power & Light Co.	BWR-MARK 1 GE 4 UE&C BRRT	2436	0754	02/07/1970 12/27/1974 11/03/1975 12/27/2014	94.1 78.3 91.7 95.4 85.8 99.0

U.S. Commercial Nuclear Power Reactors (Cont.)

Unit Operating Utility	Type NSSS AE Constructor	Licensed MWt	Net MDC	CP Issued OL Issued Comm. Op Exp. Date	1995-2000* Average Capacity Factors (Percent)
Byron 1 Exelon	PWR-DRYAMB WEST 4LP S&L CWE	3411	1105	12/31/1975 02/14/1985 09/16/1985 10/31/2024	79.5 70.6 74.0 77.6 92.0 95.7
Byron 2 Exelon	PWR-DRYAMB WEST 4LP S&L CWE	3411	1105	12/31/1975 01/30/1987 08/21/1987 11/06/2026	84.5 80.6 94.0 85.7 94.8 103.1
Callaway Union Electric Co.	PWR-DRYAMB WEST 4LP BECH DANI	3565	1171	04/16/1976 10/18/1984 12/19/1984 10/18/2024	83.7 90.0 90.9 84.8 87.2 101.1
Calvert Cliffs 1 Constellation	PWR-DRYAMB COMB CE BECH BECH	2700	0835	07/07/1969 07/31/1974 05/08/1975 07/31/2034	96.1 65.8 97.9 81.9 96.8 89.0
Calvert Cliffs 2 Constellation	PWR-DRYAMB COMB CE BECH BECH	2700	0840	07/07/1969 11/30/1976 04/01/1977 08/13/2036	80.3 98.2 81.2 97.7 86.6 100.8
Catawba 1 Duke Power Co.	PWR-ICECND WEST 4LP DUKE DUKE	3411	1129	08/07/1975 01/17/1985 06/29/1985 12/06/2024	88.2 63.6 92.8 88.2 91.7 90.0

U.S. Commercial Nuclear Power Reactors (Cont.)

Unit Operating Utility	Type NSSS AE Constructor	Licensed MWt	Net MDC	CP Issued OL Issued Comm. Op Exp. Date	1995-2000* Average Capacity Factors (Percent)
Catawba 2 Duke Power Co.	PWR-ICECND WEST 4LP DUKE DUKE	3411	1129	08/07/1975 05/15/1986 08/19/1986 02/24/2026	80.3 93.1 86.8 85.2 89.5 90.6
Clinton Exelon	BWR-MARK 3 GE 6 S&L BALD	2894	0930	02/24/1976 04/17/1987 11/24/1987 09/29/2026	75.0 65.0 0.0 0.0 57.7 84.3
Columbia Generating Station Energy Northwest	BWR-MARK 2 GE 5 B&R BECH	3486	1107	03/19/1973 04/13/1984 12/13/1984 12/20/2023	72.5 57.1 63.0 68.1 62.8 88.5
Comanche Peak 1 Texas Utilities Electric Co.	PWR-DRYAMB WEST 4LP G&H BRRT	3411	1150	12/19/1974 04/17/1990 08/13/1990 02/08/2030	77.5 76.8 94.1 86.2 85.4 95.2
Comanche Peak 2 Texas Utilities Electric Co.	PWR-DRYAMB WEST 4LP BECH BRRT	3411	1150	12/19/1974 04/06/1993 08/03/1993 02/02/2033	91.0 73.0 80.0 95.3 86.9 87.8
Cooper Nebraska Public Power District	BWR-MARK 1 GE 4 B&R B&R	2381	0764	06/04/1968 01/18/1974 07/01/1974 01/18/2014	61.7 94.5 81.5 75.2 97.3 70.6

U.S. Commercial Nuclear Power Reactors (Cont.)

Unit Operating Utility	Type NSSS AE Constructor	Licensed MWt	Net MDC	CP Issued OL Issued Comm. Op Exp. Date	1995-2000* Average Capacity Factors (Percent)
Crystal River 3 Florida Power Corp.	PWR-DRYAMB B&W LLP GIL JONES	2544	0818	09/25/1968 01/28/1977 03/13/1977 12/03/2016	101.0 33.6 0.0 88.2 88.9 97.2
Davis-Besse First Energy	PWR-DRYAMB B&W LLP BECH BECH	2772	0873	03/24/1971 04/22/1977 07/31/1978 04/22/2017	100.5 84.3 93.9 78.1 96.4 87.4
D.C. Cook 1 Indiana/Michigan Power Co.	PWR-ICECND WEST 4LP AEP AEP	3250	1000	03/25/1969 10/25/1974 08/28/1975 10/25/2014	61.6 95.3 51.9 0.0 0.0 1.5
D.C. Cook 2 Indiana/Michigan Power Co.	PWR-ICECND WEST 4LP AEP AEP	3411	1060	03/25/1969 12/23/1977 07/01/1978 12/23/2017	92.6 86.2 63.3 0.0 0.0 51.4
Diablo Canyon 1 Pacific Gas & Electric Co.	PWR-DRYAMB WEST 4LP PG&E PG&E	3338	1087	04/23/1968 11/02/1984 05/07/1985 09/22/2021	79.2 93.2 87.1 98.0 87.5 83.3
Diablo Canyon 2 Pacific Gas & Electric Co.	PWR-DRYAMB WEST 4LP PG&E PG&E	3411	1087	12/09/1970 08/26/1985 03/13/1986 04/26/2025	92.6 83.1 93.3 84.5 88.7 96.2

U.S. Commercial Nuclear Power Reactors (Cont.)

Unit Operating Utility	Type NSSS AE Constructor	Licensed MWt	Net MDC	CP Issued OL Issued Comm. Op Exp. Date	1995-2000* Average Capacity Factors (Percent)
Dresden 2 Exelon	BWR-MARK 1 GE 3 S&L UE&C	2527	0772	01/10/1966 02/20/1991 06/09/1970 01/10/2006	27.5 31.4 82.5 79.1 92.1 101.3
Dresden 3 Exelon	BWR-MARK 1 GE 3 S&L UE&C	2527	0773	10/14/1966 03/02/1971 11/16/1971 01/12/2011	51.2 43.4 59.5 88.2 90.6 93.7
Duane Arnold Nuclear Management Co.	BWR-MARK 1 GE 4 BECH BECH	1658	0520	06/22/1970 02/22/1974 02/01/1975 02/21/2014	82.8 86.2 91.2 82.3 80.1 97.5
Edwin I. Hatch 1 Southern Nuclear Operating Co.	BWR-MARK 1 GE 4 BECH GPC	2558	0805	09/30/1969 10/13/1974 12/31/1975 08/06/2014	99.6 80.7 85.7 96.5 81.1 84.5
Edwin I. Hatch 2 Southern Nuclear Operating Co.	BWR-MARK 1 GE 4 BECH GPC	2558	0809	12/27/1972 06/13/1978 06/13/2018 09/05/1979	75.0 98.8 84.2 80.6 94.4 89.5
Fermi 2 Detroit Edison Co.	BWR-MARK 1 GE 4 S&L DANI	3430	0876	09/26/1972 07/15/1985 01/23/1988 03/20/2025	66.9 62.3 63.6 67.8 100.3 86.2

U.S. Commercial Nuclear Power Reactors (Cont.)

Unit Operating Utility	Type NSSS AE Constructor	Licensed MWt	Net MDC	CP Issued OL Issued Comm. Op Exp. Date	1995-2000* Average Capacity Factors (Percent)
Fort Calhoun Omaha Public Power District	PWR-DRYAMB COMB CE GHDR GHDR	1500	0478	06/07/1968 08/09/1973 09/26/1973 08/09/2013	80.4 74.5 91.2 77.8 85.6 92.8
Ginna Rochester Gas & Electric Corp.	PWR-DRYAMB WEST 2LP GIL BECH	1520	0470	04/25/1966 12/10/1984 07/01/1970 09/18/2009	88.4 70.2 92.6 104.1 84.0 90.5
Grand Gulf 1 Entergy Operations, Inc.	BWR-MARK 3 GE 6 BECH BECH	3833	1179	09/04/1974 11/01/1984 07/01/1985 06/16/2022	79.2 89.3 102.9 82.0 79.9 100.6
H.B. Robinson 2 Carolina Power & Light Co.	PWR-DRYAMB WEST 3LP EBSO EBSO	2300	0683	04/13/1967 09/23/1970 03/07/1971 07/31/2010	86.1 91.0 103.6 87.9 95.0 104.0
Hope Creek 1 Public Service Electric & Gas Co.	BWR-MARK1 GE 4 BECH BECH	3293	1031	11/04/1974 07/25/1986 12/20/1986 04/11/2026	78.2 74.6 70.9 92.3 85.3 80.3

U.S. Commercial Nuclear Power Reactors (Cont.)

Unit Operating Utility	Type NSSS AE Constructor	Licensed MWt	Net MDC	CP Issued OL Issued Comm. Op Exp. Date	1995-2000* Average Capacity Factors (Percent)
Indian Point 2 Entergy	PWR-DRYAMB WEST 4LP UE&C WDCO	3071	0951	10/14/1966 09/28/1973 08/01/1974 09/28/2013	59.3 94.9 38.4 23.0 88.5 12.1
Indian Point 3 Entergy	PWR-DRYAMB WEST 4LP UE&C WDCO	3025	0965	08/13/1969 04/05/1976 08/30/1976 12/15/2015	17.4 69.3 51.3 89.8 86.0 99.5
James A. FitzPatrick Entergy	BWR-MARK 1 GE 4 S&W S&W	2536	0762	05/20/1970 10/17/1974 07/28/1975 10/17/2014	70.7 78.6 94.7 73.2 93.5 84.4
Joseph M. Farley 1 Southern Nuclear Operating Co.	PWR-DRYAMB WEST 3LP SSI DANI	2775	0812	08/16/1972 06/25/1977 12/01/1977 06/25/2017	80.7 100.1 75.2 78.9 97.4 71.5
Joseph M. Farley 2 Southern Nuclear Operating Co.	PWR-DRYAMB WEST 3LP SSI BECH	2775	0822	08/16/1972 03/31/1981 07/30/1981 03/31/2021	70.7 79.5 101.1 84.7 71.7 100.0
Kewaunee Nuclear Management Co.	PWR-DRYAMB WEST 2LP PSE PSE	1650	0511	08/06/1968 12/21/1973 06/16/1974 12/21/2013	84.7 70.6 52.8 78.4 98.8 82.7

U.S. Commercial Nuclear Power Reactors (Cont.)

Unit Operating Utility	Type NSSS AE Constructor	Licensed MWt	Net MDC	CP Issued OL Issued Comm. Op Exp. Date	1995-2000* Average Capacity Factors (Percent)
La Salle 1 Exelon	BWR-MARK 2 GE 5 S&L CWE	3323	1036	09/10/1973 08/13/1982 01/01/1984 05/17/2022	92.2 36.3 0.0 30.8 88.3 99.6
La Salle 2 Exelon	BWR-MARK 2 GE 5 S&L CWE	3323	1036	09/10/1973 03/23/1984 10/19/1984 12/16/2023	65.8 62.0 0.0 0.0 73.1 92.4
Limerick 1 Exelon	BWR-MARK 2 GE 4 BECH BECH	3458	1105	06/19/1974 08/08/1985 02/01/1986 10/26/2024	88.2 84.2 95.3 77.6 98.1 89.5
Limerick 2 Exelon	BWR-MARK 2 GE 4 BECH BECH	3458	1115	06/19/1974 08/25/1989 01/08/1990 06/22/2029	86.2 91.9 85.0 93.5 85.0 99.0
McGuire 1 Duke Power Co.	PWR-ICECND WEST 4LP DUKE DUKE	3411	1129	02/23/1973 07/08/1981 12/01/1981 06/12/2021	89.6 86.3 70.8 80.9 89.1 103.4
McGuire 2 Duke Power Co.	PWR-ICECND WEST 4LP DUKE DUKE	3411	1129	02/23/1973 05/27/1983 03/01/1984 03/03/2023	91.9 73.2 67.2 92.1 89.2 87.5

U.S. Commercial Nuclear Power Reactors (Cont.)

Unit Operating Utility	Type NSSS AE Constructor	Licensed MWt	Net MDC	CP Issued OL Issued Comm. Op Exp. Date	1995-2000* Average Capacity Factors (Percent)
Millstone 2 Dominion	PWR-DRYAMB COMB CE BECH BECH	2700	0871	12/11/1970 09/26/1975 12/26/1975 07/31/2015	35.5 13.4 0.0 0.0 57.9 81.7
Millstone 3 Dominion	PWR-DRYSUB WEST 4LP S&W S&W	3411	1137	08/09/1974 01/31/1986 04/23/1986 11/25/2025	80.2 24.3 0.0 34.0 82.7 99.9
Monticello Nuclear Management Co.	BWR-MARK 1 GE 3 BECH BECH	1670	0544	06/19/1967 01/09/1981 06/30/1971 09/08/2010	101.3 81.6 76.8 82.4 91.8 83.6
Nine Mile Point 1 Constellation	BWR-MARK 1 GE 2 NIAG S&W	1850	0565	04/12/1965 12/26/1974 12/01/1969 08/22/2009	87.0 94.2 54.5 87.9 72.0 94.3
Nine Mile Point 2 Constellation	BWR-MARK 2 GE5 S&W S&W	3467	1105	06/24/1974 07/02/1987 03/11/1988 10/31/2026	78.1 89.6 91.7 71.4 89.3 81.1
North Anna 1 Dominion	PWR-DRYSUB WEST 3LP S&W S&W	2893	0893	02/19/1971 04/01/1978 06/06/1978 04/01/2018	99.8 88.5 91.5 90.5 103.8 92.0

U.S. Commercial Nuclear Power Reactors (Cont.)

Unit Operating Utility	Type NSSS AE Constructor	Licensed MWt	Net MDC	CP Issued OL Issued Comm. Op Exp. Date	1995-2000* Average Capacity Factors (Percent)
North Anna 2 Dominion	PWR-DRYSUB WEST 3LP S&W S&W	2893	0897	02/19/1971 08/21/1980 12/14/1980 08/21/2020	77.2 77.7 99.7 89.0 91.4 101.8
Oconee 1 Duke Power Co.	PWR-DRYAMB B&W LLP DBDB DUKE	2568	0846	11/06/1967 02/06/1973 07/15/1973 02/06/2033	85.8 74.8 43.0 77.1 83.8 84.9
Oconee 2 Duke Power Co.	PWR-DRYAMB B&W LLP DBDB DUKE	2568	0846	11/06/1967 10/06/1973 09/09/1974 10/06/2033	94.1 59.4 79.2 72.1 84.4 100.9
Oconee 3 Duke Power Co.	PWR-DRYAMB B&W LLP DBDB DUKE	2568	0846	11/06/1967 07/19/1974 12/16/1974 12/16/2034	87.3 73.3 62.7 79.8 99.4 88.5
Oyster Creek Amergen	BWR-MARK 1 GE 2 B&R B&R	1930	0619	12/15/1964 07/02/1991 12/01/1969 12/15/2009	95.8 79.8 93.6 74.3 99.4 71.9
Palisades Nuclear Management Co.	PWR-DRYAMB COMB CE BECH BECH	2530	0730	03/14/1967 02/21/1991 12/31/1971 03/14/2007	76.0 82.9 90.8 80.0 80.2 89.6

U.S. Commercial Nuclear Power Reactors (Cont.)

Unit Operating Utility	Type NSSS AE Constructor	Licensed MWt	Net MDC	CP Issued OL Issued Comm. Op Exp. Date	1995-2000* Average Capacity Factors (Percent)
Palo Verde 1 Arizona Public Service Co.	PWR-DRYAMB COMB CE80 BECH BECH	3800	1227	05/25/1976 06/01/1985 01/28/1986 12/31/2024	79.3 80.8 98.6 87.4 88.7 100.4
Palo Verde 2 Arizona Public Service Co.	PWR-DRYAMB COMB CE80 BECH BECH	3876	1227	05/25/1976 04/24/1986 09/19/1986 12/09/2025	84.4 86.7 85.6 101.8 90.0 87.2
Palo Verde 3 Arizona Public Service Co.	PWR-DRYAMB COMB CE80 BECH BECH	3876	1230	05/25/1976 11/25/1987 01/08/1988 03/25/2027	87.1 99.9 86.5 87.6 100.3 90.3
Peach Bottom 2 Exelon	BWR-MARK 1 GE 4 BECH BECH	3458	1093	01/31/1968 12/14/1973 07/05/1974 08/08/2013	97.8 79.8 100.0 75.9 98.8 88.8
Peach Bottom 3 Exelon	BWR-MARK 1 GE 4 BECH BECH	3458	1093	01/31/1968 07/02/1974 12/23/1974 07/02/2014	78.0 98.2 79.0 90.1 89.4 99.5
Perry 1 First Energy	BWR-MARK 3 GE 6 GIL KAIS	3579	1160	05/03/1977 11/13/1986 11/18/1987 03/18/2026	89.2 73.1 80.2 96.7 89.8 93.9

U.S. Commercial Nuclear Power Reactors (Cont.)

Unit Operating Utility	Type NSSS AE Constructor	Licensed MWt	Net MDC	CP Issued OL Issued Comm. Op Exp. Date	1995-2000* Average Capacity Factors (Percent)
Pilgrim 1 Entergy	BWR-MARK 1 GE 3 BECH BECH	1998	0670	08/26/1968 09/15/1972 12/01/1972 06/08/2012	76.4 90.5 73.4 96.9 76.2 93.7
Point Beach 1 Nuclear Management Co.	PWR-DRYAMB WEST 2LP BECH BECH	1519	0485	07/19/1967 10/05/1970 12/21/1970 10/05/2010	89.3 97.7 19.4 54.9 78.4 92.3
Point Beach 2 Nuclear Management Co.	PWR-DRYAMB WEST 2LP BECH BECH	1519	0485	07/25/1968 03/08/1973 10/01/1972 03/08/2013	79.7 69.2 19.0 77.5 80.0 78.4
Prairie Island 1 Nuclear Management Co.	PWR-DRYAMB WEST 2LP FLUR NSP	1650	0513	06/25/1968 04/05/1974 12/16/1973 08/09/2013	100.6 83.0 78.4 89.7 89.0 98.9
Prairie Island 2 Nuclear Management Co.	PWR-DRYAMB WEST 2LP FLUR NSP	1650	0512	06/25/1968 10/29/1974 12/21/1974 10/29/2014	88.5 99.7 81.2 78.6 100.5 91.1
Quad Cities 1 Exelon	BWR-MARK 1 GE 3 S&L UE&C	2511	0769	02/15/1967 12/14/1972 02/18/1973 12/14/2012	87.4 39.7 82.6 42.1 94.1 91.3

U.S. Commercial Nuclear Power Reactors (Cont.)

Unit Operating Utility	Type NSSS AE Constructor	Licensed MWt	Net MDC	CP Issued OL Issued Comm. Op Exp. Date	1995-2000* Average Capacity Factors (Percent)
Quad Cities 2 Exelon	BWR-MARK 1 GE 3 S&L UE&C	2511	0769	02/15/1967 12/14/1972 03/10/1973 12/14/2012	37.1 69.1 39.0 50.6 97.9 92.1
River Bend 1 Entergy Operations, Inc	BWR-MARK 3 GE 6 S&W S&W	2894	0936	03/25/1977 11/20/1985 06/16/1986 08/29/2025	96.7 83.4 83.2 95.1 69.6 89.4
Salem 1 Public Service Electric & Gas Co.	PWR-DRYAMB WEST 4LP PUBS UE&C	3411	1122	09/25/1968 12/01/1976 06/30/1977 08/13/2016	26.0 0.0 0.0 63.1 82.7 92.2
Salem 2 Public Service Electric & Gas Co.	PWR-DRYAMB WEST 4LP PUBS UE&C	3411	1122	09/25/1968 05/20/1981 10/13/1981 04/18/2020	20.8 0.0 25.5 80.9 82.0 86.3
San Onofre 2 Southern California Edison Co.	PWR-DRYAMB COMB CE BECH BECH	3390	1070	10/18/1973 09/07/1982 08/08/1983 10/18/2013	69.1 91.0 70.5 89.1 87.9 90.7
San Onofre 3 Southern California Edison Co.	PWR-DRYAMB COMB CE BECH BECH	3390	1080	10/18/1973 09/16/1983 04/01/1984 10/18/2013	79.3 93.2 72.1 95.8 88.9 101.6

U.S. Commercial Nuclear Power Reactors (Cont.)

Unit Operating Utility	Type NSSS AE Constructor	Licensed MWt	Net MDC	CP Issued OL Issued Comm. Op Exp. Date	1995-2000* Average Capacity Factors (Percent)
Seabrook 1 North Atlantic Energy Service Corp.	PWR-DRYAMB WEST 4LP UE&C UE&C	3411	1158	07/07/1976 03/15/1990 08/19/1990 10/17/2026	83.1 96.8 78.3 81.1 85.8 78.1
Sequoyah 1 Tennessee Valley Authority	PWR-ICECND WEST 4LP TVA TVA	3411	1117	05/27/1970 09/17/1980 07/01/1981 09/17/2020	70.1 94.7 85.1 87.8 101.6 78.8
Sequoyah 2 Tennessee Valley Authority	PWR-ICECND WEST 4LP TVA TVA	3411	1117	05/27/1970 09/15/1981 06/01/1982 09/15/2021	91.7 78.3 89.2 97.3 91.8 92.3
Shearon Harris 1 Carolina Power & Light Co.	PWR-DRYAMB WEST 3LP EBSO DANI	2775	0860	01/27/1978 01/12/1987 05/02/1987 10/24/2026	79.2 93.6 78.3 93.4 96.2 91.0
South Texas Project 1 STP Nuclear Operating Co.	PWR-DRYAMB WEST 4LP BECH EBSO	3800	1251	12/22/1975 03/22/1988 08/25/1988 08/20/2027	84.9 93.1 90.1 98.4 88.0 78.2
South Texas Project 2 STP Nuclear Operating Co.	PWR-DRYAMB WEST 4LP BECH EBSO	3800	1251	12/22/1975 03/28/1989 06/19/1989 12/15/2028	90.6 95.2 91.0 90.1 89.4 96.1

U.S. Commercial Nuclear Power Reactors (Cont.)

Unit Operating Utility	Type NSSS AE Constructor	Licensed MWt	Net MDC	CP Issued OL Issued Comm. Op Exp. Date	1995-2000* Average Capacity Factors (Percent)
St. Lucie 1 Florida Power & Light Co.	PWR-DRYAMB COMB CE EBSO EBSO	2700	0839	07/01/1970 03/01/1976 12/21/1976 03/01/2016	74.9 70.9 77.8 94.9 88.9 102.0
St. Lucie 2 Florida Power & Light Co.	PWR-DRYAMB COMB CE EBSO EBSO	2700	0839	05/02/1977 06/10/1983 08/08/1983 04/06/2023	71.9 94.8 88.4 90.8 98.1 92.3
Summer South Carolina Electric & Gas Co.	PWR-DRYAMB WEST 3LP GIL DANI	2900	0945	03/21/1973 11/12/1982 01/01/1984 08/06/2022	97.5 88.0 87.5 101.8 88.2 74.9
Surry 1 Dominion	PWR-DRYSUB WEST 3LP S&W S&W	2546	0801	06/25/1968 05/25/1972 12/22/1972 05/25/2012	83.6 101.4 80.4 78.4 104.4 93.1
Surry 2 Dominion	PWR-DRYSUB WEST 3LP S&W S&W	2546	0801	06/25/1968 01/29/1973 05/01/1973 01/29/2013	80.1 86.4 91.9 100.1 83.7 92.9
Susquehanna 1 Pennsylvania Power & Light Co.	BWR-MARK 2 GE 4 BECH BECH	3441	1090	11/02/1973 11/12/1982 06/08/1983 07/17/2022	78.8 81.0 95.2 68.9 92.3 85.4

U.S. Commercial Nuclear Power Reactors (Cont.)

Unit Operating Utility	Type NSSS AE Constructor	Licensed MWt	Net MDC	CP Issued OL Issued Comm. Op Exp. Date	1995-2000* Average Capacity Factors (Percent)
Susquehanna 2 Pennsylvania Power & Light Co.	BWR-MARK 2 GE 4 BECH BECH	3441	1094	11/02/1973 06/27/1984 02/12/1985 03/23/2024	85.5 95.0 80.6 94.7 81.3 97.3
Three Mile Island 1 Amergen	PWR-DRYAMB B&W LLP GIL UE&C	2568	0786	05/18/1968 04/19/1974 09/02/1974 04/19/2014	92.8 102.8 86.0 97.7 77.4 103.5
Turkey Point 3 Florida Power & Light Co.	PWR-DRYAMB WEST 3LP BECH BECH	2300	0693	04/27/1967 07/19/1972 12/14/1972 07/19/2012	89.5 97.3 86.5 87.7 100.7 93.4
Turkey Point 4 Florida Power & Light Co.	PWR-DRYAMB WEST 3LP BECH BECH	2300	0693	04/27/1967 04/10/1973 09/07/1973 04/10/2013	99.5 87.7 89.7 101.7 94.5 91.9
Vermont Yankee VT Yankee Nuclear Power Corp.	BWR-MARK 1 GE 4 EBSO EBSO	1593	0510	12/11/1967 02/28/1973 11/30/1972 03/21/2012	86.7 84.8 95.5 71.9 90.9 101.5
Vogtle 1 Southern Nuclear Operating Co.	PWR-DRYAMB WEST 4LP SBEC GPC	3565	1162	06/28/1974 03/16/1987 06/01/1987 01/16/2027	98.1 79.8 81.2 99.6 93.5 91.2

U.S. Commercial Nuclear Power Reactors (Cont.)

Unit Operating Utility	Type NSSS AE Constructor	Licensed MWt	Net MDC	CP Issued OL Issued Comm. Op Exp. Date	1995-2000* Average Capacity Factors (Percent)
Vogtle 2 Southern Nuclear Operating Co.	PWR-DRYAMB WEST 4LP SBEC GPC	3565	1162	06/28/1974 03/31/1989 05/20/1989 02/09/2029	90.0 88.5 101.3 80.2 87.0 102.4
Waterford 3 Entergy Operations, Inc.	PWR-DRYAMB COMB CE EBSO EBSO	3390	1104	11/14/1974 03/16/1985 09/24/1985 12/18/2024	82.4 94.5 71.4 89.3 79.0 89.8
Watts Bar 1 Tennessee Valley Authority	PWR-ICECND WEST 4LP TVA TVA	3411	1118	01/23/1973 02/07/1996 05/27/1996 11/09/2035	- 89.1 77.7 94.7 84.4 92.4
Wolf Creek 1 Wolf Creek Nuclear Operating Corp.	PWR-DRYAMB WEST 4LP BECH DANI	3565	1163	05/31/1977 06/04/1985 09/03/1985 03/11/2025	98.7 80.2 82.7 101.5 89.3 88.3

*Note: Average capacity factors are listed in year order starting with 1995.

Source: NRC and licensee data as compiled by the NRC.

Abbreviations Used In Annex 1

ABB-CE	Asea Brown Boveri-Combustion Engineering		
ACE	ACEOWEN, Ateliers de Constructions Electriques de Charleroi S.A. (ACEC) and Cocerill Ougree-Providence (COP); with Westinghouse (Belgium)		
ACLF	ACECO/Creusot-loire/Framatome/Westinghouse-Europe		
AE	Architect-Engineer		
AEC	Atomic Energy Commission		
AECL	Atomic Energy of Canada, Ltd.		
AEE	Atomenergoexport		
AEP	American Electric Power		
AGN	Aerojet-General Nucleonics		
ASEA	Asea Brown Boveri-Asea Atom		
B&R	Burns & Roe		
B&W	Babcock & Wilcox		
BALD	Baldwin Associates		
BECH	Bechtel		
BRRT	Brown & Root		
BWR	Boiling-Water Reactor		
COMB	Combustion Engineering		
COMM. OP.	Date of Commercial Operation		
CON TYPE	Containment Type		
		DRYAMB	Dry, Ambient Pressure
		DRYSUB	Dry, Subatmospheric
		HTG	High-Temperature Gas-Cooled
		ICECND	Wet, Ice Condenser
		LMFB	Liquid Metal Fast Breeder
		MARK 1	Wet, Mark I
		MARK 2	Wet, Mark II
		MARK 3	Wet, Mark III
		OCM	Organic Cooled & Moderated
		PTHW	Pressure Tube, Heavy Water

Abbreviations Used In Annex 1 (Cont.)

		SCF	Sodium Cooled, Fast
		SCGM	Sodium Cooled, Graphite Moderated
CP	Construction Permit		
CP ISSUED	Date of Construction Permit Issuance		
CPPR	Construction Permit Power Reactor		
CWE	Commonwealth Edison Company		
CX	Critical Assembly		
DANI	Daniel International		
DBDB	Duke & Bechtel		
DER	Design Electric Rating		
DOE	Department of Energy		
DPR	Demonstration Power Reactor		
DUKE	Duke Power Company		
EBSO	Ebasco		
EXP. DATE	Expiration Date of Operating License		
FENOC	FirstEnergy Nuclear Operating Co.		
FRAM	Framatome		
FLUR	Fluor Pioneer		
G&H	Gibbs & Hill		
GCR	Gas-Cooled Reactor		
GE	General Electric		
GHDR	Gibbs & Hill & Durham & Richardson		
GIL	Gilbert Associates		
GPC	Georgia Power Company		
HIT	Hitachi		
HTG	High-Temperature Gas-Cooled		
HWR	Pressurized Heavy-Water Reactor		
IES	Iowa Electric		
JONES	J. A. Jones		

Abbreviations Used In Annex 1 (Cont.)

KAIS	Kaiser Engineers	
KWU	Kraftwerk Union, Siemens AG	
LIC. TYPE:	License Type	
	CP	Construction Permit
	OL-FP	Operating License-Full Power
	OL-LP	Operating License-Low Power
MAE	Ministry of Atomic Energy, Russian Federation	
MDC	Maximum Dependable Capacity - Net	
MHI	Mitsubishi Heavy Industries, Ltd.	
MWe	Megawatts Electrical	
MWt	Megawatts Thermal	
NIAG	Niagara Mohawk Power Corporation	
NPF	Nuclear Power Facility	
NSP	Northern States Power Company	
NSSS	Nuclear Steam System Supplier & Design Type	
	1	GE Type 1
	2	GE Type 2
	3	GE Type 3
	4	GE Type 4
	5	GE Type 5
	6	GE Type 6
	2LP	Westinghouse Two-Loop
	3LP	Westinghouse Three-Loop
	4LP	Westinghouse Four-Loop
	CE	Combustion Engineering
	CE80	CE Standard Design
	LLP	B&W Lowered Loop
	RLP	B&W Raised Loop

Abbreviations Used In Annex 1 (Cont.)

OL	Operating License
OL ISSUED	Date of Latest Full Power Operating License
PECO	Philadelphia Energy Company
PG&E	Pacific Gas & Electric Company
PHWR	Pressurized Heavy-Water-Moderated Reactor
PSE	Pioneer Services & Engineering
PUBS	Public Service Electric & Gas Company
PWR	Pressurized-Water Reactor
R	Research
S&L	Sargent & Lundy
S&W	Stone & Webster
SBEC	Southern Services & Bechtel
SSI	Southern Services Incorporated
STP	South Texas Project
TXU	Texas Utilities
TNPG	The Nuclear Power Group
TOSH	Toshiba
TR	Test Reactor
TVA	Tennessee Valley Authority
UE&C	United Engineers & Constructors
UTR	Universal Training Reactor
VT	Vermont
WDCO	Westinghouse Development Corporation
WEST	Westinghouse Electric

ANNEX 2: Tables 3.1 - 3.6

Table 3.1: At-power Precursors Involving Initiating Events Sorted by Event Identifier

Table 3.2: At-power Precursors Involving Unavailabilities Sorted by Event Identifier

Table 3.3: At-power Precursors Involving Initiating Events Sorted by CCDP (Conditional Core Damage Probability

Table 3.4: At-power Precursors Involving Unavailabilities Sorted by Importance

Table 3.5: Index of "Containment-Related" Events"

Table 3.6: Index of "Interesting" Events"

SOURCE: NUREG/CR-4674, Vol. 27, "Precursors to Potential Severe Core Damage Accidents: 1999-- A Status Report, July 2000

RELEVANT ARTICLE: 6

ACRONYMS:

BWR	Boiling water reactor
CCDP	Conditional core damage probability
CDP	Core damage probability
CCW	Component cooling water
ECCS	Emergency core cooling system
EDG	Emergency diesel generator
LER	Licensee event report
LOOP	Loss of offsite power
PWR	Pressurized water reactor

Table 3.1 At-power Precursors Involving Initiating Events Sorted by Event Identifier

Plant	Event identifier	Description	Plant type	Event date	CCDP	Event type
Davis Besse	LER 346/98-011	Manual reactor trip while recovering from a component cooling system leak and de-energizing of safety-related bus D1 and nonsafety bus D2	PWR	10/14/98	4.7×10^{-5}	Transient
Indian Point 2	LER 247/99-015	Loss of offsite power (LOOP) to safety-related buses following a reactor trip and tripping of a emergency diesel output breaker	PWR	08/31/99	4.7×10^{-5}	LOOP

Table 3.2 At-Power Precursors Involving Unavailabilities Sorted by Event Identifier

Plant	Event identifier	Description	Plant type	Event date	CCDP	Importance (CCDP – CDP)	Event type
Oconee 1	LER 269-99-001	Postulated high-energy line leaks or breaks in turbine building leading to failure of safety-related 4 kV switchgear	PWR	2/24/99		1.0×10^{-3} to 1.0×10^{-5}	Unavailability
Oconee 2	LER 269-99-001	Postulated high-energy line leaks or breaks in turbine building leading to failure of safety-related 4 kV switchgear	PWR	2/24/99		2.7×10^{-4} to 2.7×10^{-6}	Unavailability
Oconee 3	LER 269-99-001	Postulated high-energy line leaks or breaks in turbine building leading to failure of safety-related 4 kV switchgear	PWR	2/24/99		2.6×10^{-4} to 2.6×10^{-6}	Unavailability
Cook 1	Inspection reports: 50-315/316/97-024 50-315/316/99-010	Lack of procedure for manually backwashinging the emergency service water pump discharge strainers	PWR	12/97 5/99		3.2×10^{-5}	Unavailability
Cook 2	Inspection reports: 50-315/316/97-024 50-315/316/99-010	Lack of procedure for manually backwashing the emergency service water pump discharge strainers	PWR	12/97 5/99		3.2×10^{-5}	Unavailability
Cook 1	LER 315/99-026	High energy line break programmatic results in unanalyzed conditions	PWR	10/22/99		4.3×10^{-4}	Unavailability
Cook 2	LER 315/99-026	High energy line break programmatic results in unanalyzed conditions	PWR	10/22/99		4.3×10^{-4}	Unavailability
Cook 1	LER 315/99-031	Valves required to operate post-accident could fail to open due to pressure locking/thermal binding	PWR	12/30/99		3.7×10^{-5}	Unavailability

Cook 2	LER 315/99-031	Valves required to operate post-accident could fail to open due to pressure locking/thermal binding	PWR	12/30/99		3.7×10^{-5}	Unavailability

Table 3.3 At-Power Precursors Involving Initiating Events Sorted by CCDP

CCDP	Plant	Plant Type	Event Identifier	Description	Event Date	Event Type
5.6×10^{-4}	Davis-Besse 1	PWR	LER 346/98-006	A tornado touchdown causes a complete (weather-related) loss of offsite power (LOOP)	06/24/98	LOOP
1.4×10^{-5}	Davis-Besse 1	PWR	LER 346/98-011	Manual reactor trip while recovering from a component cooling system leak and de-energizing of safety-related bus D1 and nonsafety bus D2	10/14/98	TRANS

Table 3.4 At-Power Precursors Involving Unavailabilities Sorted by Importance

Importance (CCDP –CDP)	CCDP	Plant	Plant Type	Event Identifier	Description	Event Date	Event Type
1.1×10^{-5}	6.9×10^{-5}	Big Rock Point	BWR	LER 155/98-001	Standby Liquid Control System unavailable for 13 years	07/14/98	Unavailability
8.1×10^{-6}	9.0×10^{-6}	Byron 1	PWR	LER 454/98-018	Long-term unavailability (18 d) of an emergency diesel generator (EDG)	09/12/98	Unavailability
7.2×10^{-6}	4.6×10^{-5}	San Onofre 2	PWR	LER 361/98-003	Inoperable sump recirculation valve	02/05/98	Unavailability
2.7×10^{-6}	5.9×10^{-5}	Cook 2	PWR	LER 316/98-005	A postulated crack in a unit 2 main steam line may degrade the ability of the adjacent CCW pumps to perform their function	01/22/98	Unavailability
1.7×10^{-6}	2.0×10^{-5}	Oconee 1	PWR	LER 269/98-004, -005	Calibration and calculational errors compromise ECCS transfer to emergency sump	02/12/98	Unavailability
1.7×10^{-6}	2.0×10^{-5}	Oconee 2	PWR	LER 269/98-004, -005	Calibration and calculational errors compromise ECCS transfer to emergency sump	02/12/98	Unavailability

1.4×10^{-6}	1.9×10^{-5}	Oconee 3	PWR	LER 269/98-004, -005	Calibration and calculational errors compromise emergency core cooling system transfer to emergency sump	02/12/98	Unavailability

Table 3.5 Index of "Containment-Related" Events

Plant	Plant Type	Event Identifier	Description	Event Date	Event Type
Millstone 2	PWR	LER 336/98-011	Feedwater valves may not be able to close fully on an isolation signal	05/19/98	Unavailability

Table 3.6 Index of "Interesting" Events

Plant	Plant Type	Event Identifier	Description	Event Date	Event Type
Point Beach 1	PWR	LER 226/98-002	Loss of the station auxiliary transformer while at power	01/08/98	Transient
Columbia	BWR	LER 397/98-011	Flooding of pump room of emergency core cooling system because of a fire protection system pipe break	06/17/98	Unavailability

ANNEX 3: Figure 3.1, Total number of precursors by year

SOURCE: NUREG/CR-4674, Vol. 27, "Precursors to Potential Severe Core Damage Accidents: 1999-- A Status Report, July 2000

RELEVANT ARTICLE: 6

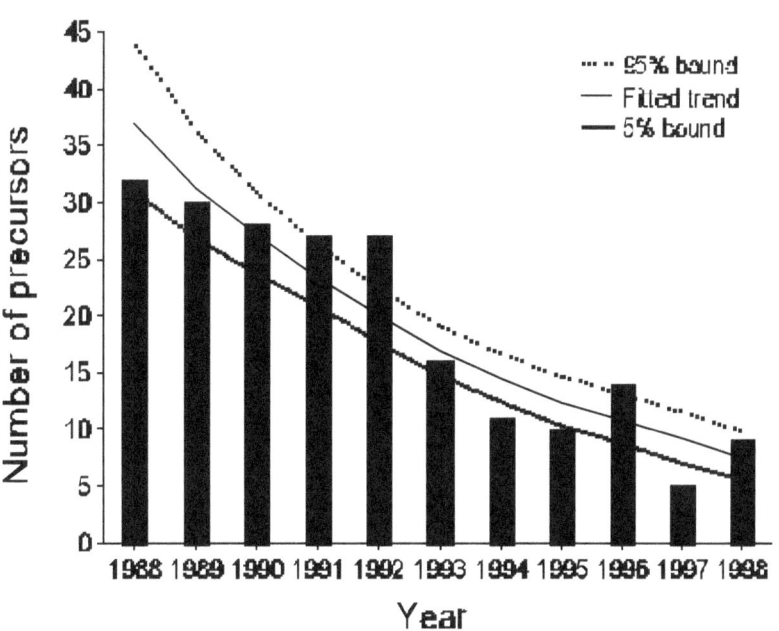

Figure 3.1 Total number of precursors by year